Using Children's Literature to Teach Problem Solving in Math

Learn how to use children's literature to engage students in mathematical problem solving. Teaching with children's literature helps build a positive math environment, encourages students to think abstractly, shows students the real-world purposes of math, builds content-area literacy, and appeals to students with different learning styles and preferences. This practical book provides specific children's book ideas and standards-based lessons that you can use to bring math alive in your own classroom.

Special Features:

- Step-by-step ideas for using children's literature to teach lessons based on the Common Core Standards for Mathematical Content in kindergarten, first, and second grade
- Scripting, modeling, and discussion prompts for each lesson
- Information on alignment to the Standards for Mathematical Practice and how to put them into student-friendly language
- Reference to a wide variety of specific children's literature that can provide a context for young children learning to engage in the standards
- Differentiated activities for students who are early, developing, and advanced problem solvers

Dr. Jeanne White teaches the math methods courses for the teacher candidates in the early childhood, elementary and special education programs at Elmhurst College. She supervises student teachers in the early childhood and elementary programs and works with in-service teachers in the Master of Education in Teacher Leadership graduate program.

Using Children's Literature to Teach Problem Solving in Math

Addressing the Common Core in K–2

Jeanne White

Routledge
Taylor & Francis Group

NEW YORK AND LONDON

First published 2014
by Routledge
711 Third Avenue, New York, NY 10017

and by Routledge
2 Park Square, Milton Park, Abingdon, Oxon OX14 4RN

Routledge is an imprint of the Taylor & Francis Group, an informa business

Library of Congress Cataloging-in-Publication Data
White, Jeanne. Using children's literature to teach problem solving in math:
addressing the common core in K-2/Jeanne White.
  pages cm
  Includes bibliographical references.
  1. Mathematics—Study and teaching (Elementary)
  2. Mathematics—Study and teaching (Elementary)—Standards.
  3. Language arts (Elementary)
  4. Language arts—Correlation with content subjects.
  I. Title.
  QA135.6.W479 2013
  372.7'049—dc23   2013034345

ISBN: 978-0-415-73532-2 (hbk)
ISBN: 978-0-415-73231-4 (pbk)
ISBN: 978-131-584922-5 (ebk)

Typeset in Bembo and Helvetica Neue
by Florence Production Ltd, Stoodleigh, Devon, UK

Printed and bound in the United States of America by Publishers Graphics,
LLC on sustainably sourced paper.

Contents

About the Author

Dr. Jeanne White has been an educator since 1992 when she began teaching elementary school in the south suburbs of Chicago. She earned her doctorate in Curriculum and Instruction in 2003 and began her career at Elmhurst College as part of the full-time faculty in 2005. She teaches the math methods courses for the teacher candidates in the early childhood, elementary and special education programs. She supervises student teachers in the early childhood and elementary programs and works with in-service teachers in the Master of Education in Teacher Leadership graduate program.

She has presented internationally, nationally and locally on topics of math education, specializing in Kindergarten through second grade. She has conducted workshops for teachers in Australia and South Africa on how to use everyday objects to facilitate early mathematics instruction. She does consulting for school districts on implementing the Common Core State Standards for Mathematics and has written numerous articles on math education.

Acknowledgments

I would like to thank my wonderful parents and family who have always supported everything I do and have been willing recipients of my writing projects over the years. I would like to thank my first and second grade students in Orland Park School District 135, the middle school and high school students I have tutored, and my college students in the Department of Education at Elmhurst College, all who have challenged me to think about how children learn mathematics. I have to thank my editor, Lauren Beebe, who was open to my ideas and continued to work with me to flesh out a viable concept for this book. The reviewers that Lauren chose were crucial to the writing process as well. Lastly, I would like to thank my fourth grade teacher at Field School in Harvey, Illinois, Joanne Penn, who is the reason why I wanted to become a teacher.

Introduction

At the root of the Common Core State Standards for Mathematics are the eight Standards for Mathematical Practice (SMP), describing "varieties of expertise that mathematics educators at all levels should seek to develop in their children (CCSSI, 2010, p. 6)." As students learn skills and strategies for performing calculations, they must also acquire skills and strategies for proving their reasoning, communicating, representing and making connections as they solve mathematical problems. Unlike many of the previous state standards used by school districts, the Common Core State Standards for Mathematics include students at the Kindergarten level. This provides opportunities for teachers to instil problem-solving skills at an early age, which can set the foundation for critical thinking in mathematics throughout the elementary grades.

Students in Kindergarten, first and second grade may have difficulty understanding the language of the eight SMP as they are written in the Common Core State Standards for Mathematics. Teachers can get together with colleagues within their school or within their district to rewrite each SMP in student-friendly language. One example of how the SMP can be rewritten is shown in Table 0.1.

Introducing students to mathematical problems can be challenging considering students in Kindergarten and first grade are at early stages of learning to read, write, add and subtract. One way teachers can facilitate the teaching of word problems is to use children's literature as the context. In this way, teachers can show illustrations from the book, use characters from the story and discuss any unfamiliar vocabulary before or during the presentation of the book. When teachers incorporate children's literature into mathematics, opportunities arise for young children to see math in their own lives. If they see how characters in a story use math to solve problems, they can better understand how people around them use math to solve everyday problems. The use of literature can also facilitate problem solving by presenting a common language and structure for the teacher and students to use while they engage in the math content.

Each chapter in this book is dedicated to one of the eight SMP with examples of how activities based on children's literature can be used as a way to apply essential aspects of each standard. The activities are all based on the Standards for Mathematical Content in Kindergarten, first and second grade. Teachers

TABLE 0.1 Student-Friendly Language for Standards for Mathematical Practice

1   Make sense and persevere in solving problems.

I can try many times to understand and solve a math problem.

2   Reason abstractly and quantitatively.

I can think about the math problem in my head, first.

3   Construct viable arguments and critique the reasoning of others.

I can make a plan, called a strategy, to solve the problem and discuss other children's strategies too.

4   Model with mathematics.

I can use math symbols and numbers to solve the problem.

5   Use appropriate tools strategically.

I can use math tools, pictures, drawings and objects to solve the problem.

6   Attend to precision.

I can check to see if my strategy and calculations are correct.

7   Look for and make use of structure.

I can use what I already know about math to solve the problem.

8   Look for and express regularity in repeated reasoning.

I can use a strategy that I used to solve another math problem.

Source: White and Dauksas, 2012, p. 442.

would choose the literature books and activities used for each SMP based on the type of problem solvers in their classroom—early, developing or advanced problem solvers. Teachers do not have to start with Chapter 1 but should use activities from any chapter that would be relevant for their current students. Kindergarten teachers might focus on one SMP a month, or may just choose a few to focus on for the year. First and second grade teachers could build on their students' knowledge of the SMP from previous grades, focusing on one or two each month, while helping their students connect the ideas reflected in each SMP, so they are not viewed in isolation.

Types of Problem Solvers

It is useful for teachers to think about their students as *problem solvers* and to use those words with their students during instruction. However, based on the development, prior experience and mathematical readiness of the students, teachers witness various types of problem solvers within one class. In order for teachers to create appropriate experiences during the school day for these various learners, they should identify the type of problem solvers they have in their class.

The early problem solver is one who needs a thorough explanation of the procedure and process for solving a problem. This may be the first time the student is encountering a written word problem. Teachers should model each step by using a think-aloud teaching method so the learner is aware there is a process one must go through in order to reach a solution. Students who are English Language Learners will also benefit from a great deal of modeling and simplification of word problems.

Children who are developing problem solvers have a strategy for attempting the problem, although it may not be the most efficient or appropriate strategy. These problem solvers realize a plan is needed and there is a process involved in finding the answer. They may be reliant on key words, such as "in all" or "how many more," to be able to determine which operation would be appropriate in order to solve the problem—which is a strategy that does not always prove correct.

Advanced problem solvers are able to make a plan and start solving a problem, though they may not always produce a viable solution. They can recognize the type of tools needed to solve a problem such as a diagram, ruler or calculator. They have a few strategies under their belt such as drawing a picture or simplifying the problem and can identify the appropriate operation based on the situation in the problem. These problem solvers can also explain how they arrived at their solution by explaining what they did in each step and why they did it.

Children can be characterized as each type of problem solver throughout their school career. A child may be an advanced problem solver for word problems involving addition or subtraction but then move back to a developing problem solver for multiple-step word problems. The number of examples and amount of time needed for children to move from one type to the next will vary by grade level as well as be determined by the type of learners in your classroom. You may consider providing small group instruction for children who are still at the early stage, or meeting with children individually to expose them to more practice as well as to probe their thinking. Some children may need more time to build their content knowledge before they can move into early problem solving.

Share the characteristics of each type of problem solver with your class so children can self-identify which stage they are currently exemplifying. This information can also be shared with families, resource teachers, classroom aides and teacher candidates so everyone is using the same terminology and knows how to advance children through the stages of problem solving.

Creating a Problem-Solving Community

Set the stage for problem solving in your classroom by reading excerpts from the book *The Math Curse* (1995) by Jon Scieszka. In this story, the narrator finds himself plagued by looking at every facet of his life as a math problem—eating his cereal, getting dressed and moving through his school day. Then challenge your class to think of their day as a series of math problems. What time do they have to leave the house in order to get to school on time? How many boys and girls are in the class? How many rode the bus? How many did not? Start a bulletin board for questions created by the children and choose one problem to solve each day. Share events from your life that can be put into a word problem. Communicate to your class that everyone is a problem solver, including yourself.

Decorate your classroom in a way that demonstrates the importance of mathematics to students and others who walk into your classroom. You should have a Math Word Wall that includes each of the mathematical terms used at your grade level, with a definition, picture or example. There should be posters on the wall for mathematical concepts such as shapes, fractions, money and various types of graphs. In the classroom library, there should be children's literature that includes math concepts such as the titles used in this book. Math instruction should not be restricted to a specific time of day but should be practiced throughout the day. Students can keep track of their class schedule and the time on the clock, they can count the number of hot lunches needed, they can help the physical education teacher count out groups for teams, and they can practice ordinal numbers when they line up first, second, third, etc.

Teach children how to communicate with regard to problem solving. For instance, how can they learn to listen when a peer is speaking? Students can practice with partners where one is speaking and the other is listening. One way to signal who is able to talk is by using a mouth on a stick and an ear on a stick. Give one child the mouth on a stick, which is the signal for that child to speak first, explaining the problem-solving strategy first while the child holding the ear on a stick must only listen. Then they switch roles and the other child is able to explain. Students who are English Language Learners, those who receive speech services that would inhibit others' ability to understand them when they are speaking, or students who are non-verbal, can have the ear while their partner speaks. If they are able, these learners could then draw or label their paper rather than speak when it is their turn to share.

When students are working with a partner or group to solve a problem they can also use prompts such as, "Can you speak louder?" or "I didn't understand how you solved it. Can you explain it again?" or "Draw a picture so I can see what you did." Provide sentence starters so students can begin to share their process such as, "The first thing I did was . . .," "First I drew a picture of . . .," "I knew it was addition because. . . ."

Teachers should model how students communicate with a peer and with a small group. It is not enough to simply tell your students to talk about how they found their answer. Teachers can build a community of problem solvers through the use of modeling, providing prompts and sentence starters as well as moving from simpler word problems to more complex problems, throughout the school year.

1

Make Sense and Persevere

Unpacking the Standard—SMP 1: Make Sense of Problems and Persevere in Solving Them

How many minutes would a child have to work on a word problem in order for you to determine the child has persisted in solving the problem? Do children know what is meant by perseverance? How can children persist if they do not understand the context of the problem? Making sense of a problem begins with the ability to understand the words, symbols, numbers and figures in the problem. English Language Learners may not be familiar with regional dialect and outdated terms; early readers may be overwhelmed by a long paragraph; inexperienced problem solvers may not know how to interpret symbols and figures that accompany a problem.

Children who are successful at SMP1 read the problem more than once in order to make sense of the information provided. They are able to explain the problem to a peer in their own words. They can analyze what is given, what is not given and the goal of the problem. They can draw a picture or use objects to represent the situation in the problem. They can make several attempts to find the answer, considering the strategies of other problem solvers. They continually ask themselves if their strategy and their answer make sense.

In order to fully apply SMP 1 when approaching a word problem, children should be able to take ownership of their procedures by using the following *I Can* statements:

- *I can explain the meaning of the problem in my own words.*
- *I can analyze what is given, what is not given and the goal of the problem.*
- *I can use a picture or concrete objects to understand and solve the problem.*
- *I can understand the strategies of others.*
- *I can ask myself, does this make sense?*

Early Problem Solvers

Seven Blind Mice by Ed Young—Decomposing Numbers

The book *Seven Blind Mice* (1992) represents a version of the folktale, The Blind Man and the Elephant, and follows seven blind mice who attempt to figure out the nature of a strange Something at the pond near their home. They must rely on their sense of touch to make sense of the object. Each mouse travels alone to the pond and feels a different part of the object, resulting in disagreements over the identity of the object. The last mouse to go to the pond, White Mouse, first listens to the ideas of the other mice concerning the identity of the object. She attempts a different strategy and runs across the strange Something from end to end, making sense of the various parts of the object mentioned by the other mice to reveal it is an elephant. You can use this book to illustrate SMP1 by providing a concrete example of how to make sense of a problem and then persist at finding its solution as well as incorporating how to decompose the number seven.

Operations and Algebraic Thinking K.OA

Understand addition as putting together and adding to, and understand subtraction as taking apart and taking from.

Decompose numbers less than or equal to 10 into pairs in more than one way, e.g., by using objects or drawings, and record each decomposition by a drawing or equation (e.g., $5 = 2 + 3$ and $5 = 4 + 1$).

As you read the book, engage children in a discussion by asking the following questions based on the *I Can* statements:

- What was the problem in the book and why was it a problem for them? *Explain the meaning of the problem.* (One day the mice found a strange Something at the pond and they were afraid because they didn't know what it was; they didn't know if the strange Something would drink all of the water at their pond or not let them get to the water.)

- What information could the mice use to solve their problem since they are blind? *Analyze what is given, what is not given and the goal.* (They could touch the strange Something but they couldn't see it; they could only use the information from their sense of touch to try to identify the object.)

- What were some of the ways the mice tried to solve their problem? *Use a picture or concrete objects to understand and solve the problem.* (They went one at a time to the pond to find the object; they each touched the object and then told the other mice what they thought it was; each mouse wanted to experience it for him/herself in order to figure out what it was.)

■ What did White Mouse do differently than the other mice? *Ask yourself, does this make sense?* (She wasn't sure if the answers of the other mice made sense so she wanted to go to the pond too; she decided to touch every part of the object instead of just one part and see if she could agree with one of the answers.)

■ How did she use what she learned from the other mice? *Understand the strategies of others.* (White Mouse listened to the ideas of the other mice before she went to the pond to find the object; she used descriptions given by the other mice to help form her answer.)

For a second reading of the book, model how to use the *I Can* statements to solve a word problem related to the story:

There were seven blind mice who found a strange Something by their pond then ran back home. One of the mice, Red Mouse, went back to the pond the next day to touch the strange Something. While he was at the pond, how many mice were still at home?

Use a think-aloud method to begin modeling the *I Can* statements:

"First, I have to think, *can I explain the meaning of the problem in my own words?* There are 7 mice. 1 mouse is at the pond. How many mice are still at home? Next, I have to *analyze what is given, what is not given and my goal.* I am given the total number of mice at the beginning of the problem, which is 7. I am given the number of mice that went to the pond, which is 1. I am not given the number of mice that are still home, which is my goal. I can use 7 counters to *understand and solve the problem* by acting it out. I will put 7 counters in a square, which will be the home. I will move 1 of the counters to an oval, which will be the pond. Now I can count the mice that are still at home. There are 6 mice still at home."

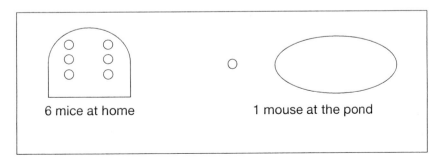

6 mice at home   1 mouse at the pond

FIGURE 1.1

Once the children have observed and listened to your think–aloud strategy, they can become more involved in applying the next *I Can* statement. Draw a picture based on your concrete model and ask the children if your picture represents the problem and if your answer makes sense while further modeling:

"I'm not finished yet because I have to look back at my answer and ask myself, *does this answer make sense*? Well, if I started out with 7 mice and 1 mouse was taken away from the group, there would be 6 mice left because 6 is one less than 7. So my answer does make sense. Does anyone have another way, or strategy, to solve the problem?"

Children can share their ideas for solving the problem with a peer using objects, pictures, words and symbols based on their readiness levels and experiences. Ask children to share their work with the class and see if others came up with the same answer but used a different strategy. If children are able, they should explain their strategy, and you can restate their explanation in order to model the last *I Can* statement:

"Janet, let me see if I understand your strategy for solving the mouse problem. You drew 7 mice then you crossed 1 out because you said that would be the mouse at the pond. Then you counted the mice that were not crossed out and you also got 6 for your answer. Okay, I *understand your strategy for solving the problem*. Thank you, Janet."

It is important to do similar problems with the class so they can generalize the procedures for making sense of a problem and trying different ways to solve it. Show the children a problem using a different number but a similar context:

There were 8 little rabbits who found a carrot patch but the farmer chased them back to their home. One of the rabbits, Brown Rabbit, went back to the patch to dig out a carrot. While he was at the carrot patch, how many rabbits were still at home?

Help the children think about how they should approach this problem by asking some questions:

- How is this problem similar to the problem about the seven blind mice? (There are animals that find something then run back home; one of the animals goes back while the others stay home; you have to find how many animals are still home.)

- How is it different? (There are rabbits instead of mice; there are 8 animals instead of 7; they find a carrot patch instead of a strange Something; they go home because the farmer chases them instead of being scared off by the strange Something; the rabbits are not blind.)

After the children have discussed similarities and differences between the two problems, use a think-aloud strategy to explain how to use the *I Can* statements with the rabbit problem:

> "I can begin to solve this problem the way I solved the problem about the 7 blind mice. First, how *can I explain the meaning of the problem in my own words*? There are 8 rabbits. 1 rabbit is at the carrot patch. How many rabbits are still at home? Next, I have to *analyze what is given, what is not given and my goal*. I am given the total number of rabbits, which is 8. I am given the number of rabbits that went to the carrot patch, which is 1. I am not given the number of rabbits that are still home, which is my goal."

Before explaining the next step, solicit ideas from the class to see if they can think of strategies to use to find the answer. You can ask guiding questions to help them:

- What can I do to find out how many rabbits are still at home?
- What strategy did I use with the problem about the seven blind mice?
- Can I use that same strategy with this problem?
- Are there other strategies I could use?

Once they have shared ideas about how you could solve the problem, choose a different strategy so they can see there are other strategies that can work:

> "Now I have to think about a strategy I can use to help me *understand and solve the problem*. For the 7 blind mice problem, I used counters and acted out the problem. That helped me see how many mice were still at home. This time I will use a different strategy. I'll use the strategy Janet shared with us. I will draw 8 circles which will be the rabbits at the beginning of the problem. Then I will cross out one of the circles because that will be the rabbit that went to the carrot patch. I will draw 1 circle over here, which is the rabbit who left home. Now I can count the rabbits that are still at home. There are 7 rabbits still at home."

7 rabbits are at home   1 rabbit is at the carrot patch

FIGURE 1.2

Developing Problem Solvers

MATH-Terpieces: The Art of Problem Solving by Greg Tang—Using 2 and 3 Addends to Find a Sum

In the book *MATH-Terpieces: The Art of Problem Solving* (2003) famous artwork serves as a backdrop for the reader to be able to add groups of objects to find sums up to a target number. Once you have modeled how to use the *I Can* statements using a few different word problems, children should practice using the *I Can* statements to solve problems with a peer as they use drawings and equations to add three whole numbers.

Operations and Algebraic Thinking 1.OA

Represent and solve problems involving addition and subtraction.

Solve word problems that call for addition of three whole numbers whose sum is less than or equal to 20, e.g., by using objects, drawings and equations with a symbol for the unknown number to represent the problem.

The first painting, Ballet Rehearsal on Stage by Edgar Degas, is displayed on the left side of the two-page spread with groups of ballet shoes on the right side. The problem posed to the reader is:

Can you make 7 with these shoes?
Three clever ways earn rave reviews!

Ask questions that can allow children to solve the problem with their peers:

- Can you *explain the meaning of the problem in your own words* to your partner? (How can I make 7 by putting together groups of shoes? There are 3 ways to do it.)

- Can you *analyze what is given, what is not given and the goal* of this problem? (There are 5 different groups of ballet shoes in the picture. There is a group of 5, a group of 2, a group of 4, a group of 3 and 1 shoe by itself. I have all of the information I need to find the 3 ways to make 7.)

- Can you *use concrete objects or pictures to help you understand and solve the problem?* You can use paper to draw the shoes or cubes snapped together to represent the groups of shoes. Keep track of all of the combinations you and your partner try, even if some of your ideas don't work. (I can count groups of shoes to see if they add up to 7. I have solved the problem when I have found 3 different ways to make 7 with the groups of shoes in the picture.)

- What are your answers? *Do they make sense?* How can you check? (I can write number sentences for my answers and check with the cubes to make sure my 3 ways add up to 7).

- What strategy did you and your partner use? Who would like to *explain their strategy* to the class? (We drew pictures of the shoes and the first one we wrote was 4 and 3 because we knew 4 plus 3 equals 7. Then we looked at the biggest group, which was 5, and we knew to put it with the 2 because we knew 5 plus 2 equals 7. Then we were stuck because there is 1 shoe left but there is no group of 6 shoes to go with it.)

- Did anyone else get only two ways to make 7? Who else got 4 + 3 and 5 + 2? Is there another way? Greg Tang said there are 3 ways to do it with the groups of shoes on the page. Do you have to use only 2 groups of shoes in your answers? Can you use a group of shoes more than once? Can you do it with 3 groups of shoes? (Yes. We found the last way. You have to use 3 groups of shoes. You can use the group of 2 shoes, the group of 4 shoes and the 1 shoe that was leftover. We know this works because we checked and 2+4+1=7.)

If the children had difficulty finding the solution with 3 addends, use a think-aloud method to explain it:

"When I was trying to solve the problem, I tried to use a system. I began with the largest group of shoes first, the group of 5, and looked for a number I could add to 5 to equal 7. I know 5 + 2 = 7 so I used the group of 2 shoes. Then I used the second largest group of shoes, which was 4, and I matched that with the group of 3 shoes because I know 4 + 3 = 7. The only shoe left was 1 by itself and there was no group of 6 shoes to add to it to equal 7, since I know 6 + 1 = 7. But then I thought there might be a way to combine two groups of shoes to make 6. If I used the group of 4 shoes and the group of 2 shoes I could make a group of 6 and that would go with the 1 shoe by itself."

These examples illustrate not only the use of the five *I Can* statements for SMP 1 but offer opportunities for children to persist in finding the less obvious way to find the sum of seven in the problem, using three addends rather than just two. Some partners may have found all three ways to make seven with little effort. Be sure to allow enough time for children to try to discover all three ways with their partner as well as time for children to share their strategies with the class. Some may use drawings, some can add cubes or other manipulatives, and some may be at the abstract level and only use the number of shoes in each group to find the sums.

Often, when children realize some of their classmates have already found the solution, they are motivated to persist and find the solution themselves. They realize a solution does exist and if their peers can find it, so can they. Other children may actually feel discouraged when their classmates find the solution quickly while they are still struggling to find it. Give children many opportunities to discuss ideas and strategies with their peers so they can see there are many ways to solve a problem. Switch up partners often so children can work with peers of various levels of mathematical competency and understanding.

Read other pages in the MATH-terpieces book so children can practice solving the problems with their partner and start using strategies shared by their peers. As they work with their partner on a few more problems, ask guiding questions such as:

- How would you explain the meaning of this problem to another person?

- What information was given and what was your goal?

- Did you use the same strategy as on the first problem? Which strategy did you use?

- Which problems were easy to solve? Which problems were tricky? Why?

- Did you have to add more than two groups of objects like we did in the first problem? Give me an example where you did this.

- How did you check to see if your answers made sense?

- Did you try strategies suggested by your classmates? Which ones?

Advanced Problem Solvers

Splash! By Ann Jonas—Representing Addition and Subtraction Problems

In the book *Splash!* (1995) a girl tells a story about her many pets and how one day some of them fell or jumped into the pond while others came out of the pond. At the bottom of each page the same question is repeated, "How many are in my pond?" Children who have had practice using the five *I Can* statements while solving problems with their peers may be ready to try solving some on their own with addition and subtraction.

Operations and Algebraic Thinking 2.OA

Represent and solve problems involving addition and subtraction.

Use addition and subtraction within 100 to solve one- and two-step word problems involving situations of adding to, taking from, putting together, taking apart and comparing, with unknowns in all positions, e.g., by using drawings and equations with a symbol for the unknown number to represent the problem.

At first, provide a checklist with the five statements so children can begin to learn them and eventually use them automatically. Provide a large piece of paper and a manipulative such as circle counters so they can draw a picture or use the concrete objects to solve the problem. Challenge the children by telling them you have a longer problem for them to solve based on a book you will soon read. You will give them time to try solving it on their own. After about 10 minutes, you will all discuss the problem and then you will give them more time to go back and try again to solve it on their own if they need it. Display the following problem:

I have a pond in my backyard.

I have one turtle, two catfish, three frogs and four goldfish.

All of my fish are in the pond.

The turtle jumps into the pond.

One frog jumps in. My cat falls in. My dog falls in.

How many are in my pond?

As they are working, walk around to see what they are drawing on their paper or modeling with their counters. Encourage the children to think of as many different strategies as they can to find the answer. Let them know they should not say the answer if they think they have found it but just write it on the top of their paper and circle it. After most children have attempted to solve the problem, stop them and ask guiding questions for the discussion:

- Can you *explain the meaning of the problem in your own words?*
- What *information is given, what is not given and what is your goal?*
- What strategies did you use when you started to solve the problem? Did you *use a picture or concrete objects to understand and solve the problem?*
- Why did you choose that strategy?

If there were many who did not get the correct answer, did not finish solving the problem, or were having difficulty choosing an appropriate strategy, allow children to talk with their peers about what they have drawn or represented, then provide 5–10 more minutes to work on the problem. Walk around again to see if they are using the information from their discussion to solve the problem. When most have provided an answer, stop and ask:

- Did anyone try to *understand the strategies of others* in our class? Which strategies?
- How did you *check your answer to see if it made sense?*

Then read the book and talk about how the illustrations can provide the answers to the question on each page because we can count the animals in the pond.

Let them know this is another strategy to solve the problem. As students become independent problem solvers, provide them with questions they can use to help them make sense of a problem, perhaps in the form of a poster hung in the class:

- What is the setting or situation in the problem?
- Would I be able to draw a picture to show what is happening in the problem?
- Do I understand the words, symbols and numbers in the problem?
- Are there any words I don't know? Are there illustrations or other words in the problem I can use to help me figure out unknown words?
- Can I use a strategy I used on another problem?
- Can I use a strategy suggested by another student?

Concluding Remarks

Children have to learn to be patient in their quest for perseverance. They should learn it is okay to put aside a problem and come back to it later like they did in the problem from the book *Splash!* Often a solution emerges when our brains are concentrating on another task or when we hear how others are thinking about the same problem. It is with experience children will learn more strategies they can apply to a problem if they are unable to solve it on their first attempt.

2

Reason Abstractly

Unpacking the Standard—SMP2: Reason Abstractly and Quantitatively

Children are exposed to problem solving in Kindergarten in a concrete way, with physical objects such as toys or blocks, or acting out a problem-solving situation. When children are in first and second grade, they gradually move away from the concrete stage of learning and are exposed to number models representing the problem-solving situations. Exposing children to this connecting stage, in which there are still physical or visual representations alongside the numbers and symbols, is critical so children will be able to successfully reach the abstract stage. It is also crucial for problem solvers to be able to recognize and use the properties of operations when applying SMP2.

Children who are successful at SMP2 can see relationships among the numbers so they can make decisions about which operation would be most efficient to solve the problem. They can decontextualize the situation by representing it symbolically ($5 + ? = 10$), but they must also contextualize the situation by examining the referents (objects, people, etc.) in the problem in order to determine which symbols to use. Through experience with various situations, children can determine whether the situation involves a group added to another group, if there are groups being compared, if there are groups taken from a larger group, as well as which operation is represented by these actions.

In order to fully apply SMP2 when approaching a word problem, children should be able to take ownership of their procedures by using the following *I Can* statements:

- *I can understand how the numbers in the problem are related.*
- *I can use the units in the problem.*
- *I can use properties of operations.*
- *I can represent the problem with symbols.*

Early Problem Solvers

Ten Flashing Fireflies by Philomen Sturges—Exploring the Commutative Property of Addition

It is important in the early grades to expose children to the Commutative Property of Addition and to use this term, rather than refer to addition facts that follow this property as *turnaround facts, flip-flop facts,* or other simplified terms. In the book *Ten Flashing Fireflies* (1995) a young boy and girl are outside catching fireflies in a jar. As the children catch the fireflies one by one, the fireflies captured in the jar are shown on the left and the remaining fireflies in the sky are shown on the right of the two-page spread. This provides an opportunity to explore the Commutative Property of Addition as well as practice the addition facts for 10.

Operations and Algebraic Thinking 1.OA

Understand and apply properties of operations and the relationship between addition and subtraction.

Apply properties of operations as strategies to add and subtract. *Examples: If 8 + 3 = 11 is known, then 3 + 8 = 11 is also known. (Commutative Property of Addition.)*

As you read the book, your class can keep track of the number of fireflies in the jar and in the sky by creating a table to organize the information:

TABLE 2.1

Fireflies in the jar	Fireflies in the sky	Total number of fireflies	Equations
0	10	10	0 + 10 = 10
1	9	10	1 + 9 = 10
2	8	10	2 + 8 = 10
3	7	10	3 + 7 = 10
4	6	10	4 + 6 = 10
5	5	10	5 + 5 = 10
6	4	10	6 + 4 = 10
7	3	10	7 + 3 = 10
8	2	10	8 + 2 = 10
9	1	10	9 + 1 = 10
10	0	10	10 + 0 = 10

After reading the story, ask the class some questions about the information in the table:

- What do you notice about the numbers in each row? (They add up to 10.)
- What do you notice about the numbers in the first column? (They are in counting order from 1 to 10.)
- What do you notice about the numbers in the second column? (They are in backward counting order from 10 to 0.)
- Let's look at the column labeled Equations. How is each equation created from the numbers in that row? (The equation is the first number plus the second number and it equals 10.)
- Let's write the first equation and the last equation next to each other: 0 + 10 = 10, 10 + 0 = 10. Do you notice any similarities or differences in these two equations? (They both have the same numbers—0 and 10. They both equal 10. They have the numbers in a different order. One has 0 first then 10. The other one has 10 first then 0.)
- Let's look for this relationship in other sets of equations in this column. I want you to write down as many as you can find. (Call on students to share their answers, asking each child to tell you how the two number models are similar and how they are different, based on your example.)
- There is a special name for math facts that use the same two addends but in a different order, such as these. It is called the Commutative Property of Addition. It makes it easier to learn your math facts because if you know one fact, such as 8 + 2 = 10, then you know that 2 + 8 will also equal 10.

Rooster's Off to See the World by Eric Carle—Exploring the Associative Property of Addition

After practicing the Commutative Property of Addition, your students will be ready to explore the Associative Property of Addition with the book *Rooster's Off to See the World* (1972). In this book, one rooster sets off to see the world and encounters several animals on his journey. There is a pattern to the number of animals who join the rooster; in the top right corner of each page a small graphic is shown representing each set of animals as they appear in the story, and then each set of graphics disappears as the animals change their mind and return home. You can also review the Commutative Property before you introduce the Associative Property.

Operations and Algebraic Thinking 1.OA

Understand and apply properties of operations and the relationship between addition and subtraction.

Apply properties of operations as strategies to add and subtract.
Examples: If 8 + 3 = 11 is known, then 3 + 8 = 11 is also known. (Commutative Property of Addition.) To add 2 + 6 + 4, the second two numbers can be added to make a ten, so 2 + 6 + 4 = 2 + 10 = 12. (Associative Property of Addition.)

As you read the book, create a similar table to keep track of the number of animals who join the rooster:

TABLE 2.2

Rooster	Cats	Frogs	Turtles	Fish	Equations
1 rooster					1 + 0 = 1
1 rooster	2 cats				1 + 2 = 3
1 rooster	2 cats	3 frogs			1 + 2 + 3 = 6
1 rooster	2 cats	3 frogs	4 turtles		1 + 2 + 3 + 4 = 10
1 rooster	2 cats	3 frogs	4 turtles	5 fish	1 + 2 + 3 + 4 + 5 = 15

After reading the story, ask the class some questions about the information in the table:

■ How did the number of animals change in each row? (There was only 1 animal in the first row. Then there were 2 more animals, then 3 more animals, then 4 more animals, then 5 more animals.)

■ How did our equation change in each row? (First there was only 1 animal then there were 3, then 6, then 10, then 15. We had to keep adding another number but we started with 1 each time.)

■ Let's look at the first equation, 1 + 0 = 1. What is 0 + 1? How do you know? (It is also 1. It is the same answer as 1 + 0.) Do you remember what this property is called, when we change the order of the two addends but we get the same answer? (Commutative Property of Addition.)

■ Can we use the Commutative Property with the second equation? (Yes. We can write 2 + 1 = 3.)

Pass out resealable sandwich bags of 15 squares, with a picture of each animal on a square, corresponding to the number of each animal in the story. Have the students take out the square with the rooster, the two squares with a cat on each and the three squares with a frog on each:

■ Let's use our pictures of animals to represent the third equation, 1 + 2 + 3 = 6. We are going to add the first two animals, 1 rooster and 2 cats. How many animals do we have? (We have 3 animals.)

■ Now we will add the three frogs. How many animals do we have now? (We have 6 animals.)

■ Let's try it another way. Let's start with the 2 cats and add the 3 frogs. How many animals do we have? (We have 5 animals.) Now we will add the 1 rooster. How many animals do we have now? (We have 6 animals.)

■ Did it matter which order we added the animals? Would we get the same answer if we started with the 1 rooster, added the 3 frogs and then added the 2 cats? Let's try it.

■ So we found out that we will get the same answer if we add the numbers in a different order. This is called the Associative Property of Addition. You can use this property when there are more than two addends because you might find it easier to add them in a different order.

If your students are able to apply this property with three addends, use the rest of the table to have them make new equations with four and five addends. Encourage them to find an easier way to add multiple addends such as making a ten or starting with the larger number.

Developing Problem Solvers

Each Orange Had 8 Slices by Paul Giganti, Jr.—Demonstrating Fluency for Addition

Now that your students are learning some strategies for adding numbers, they will be eager to use them with another book. There are opportunities in *Each Orange Had 8 Slices* (1992) for children to count, add and even multiply if they are ready for it. Each page illustrates a group of objects with an equal number of smaller objects and questions for the reader to answer, providing opportunities for your students to think flexibly as they develop strategies for adding sets of numbers.

Operations and Algebraic Thinking 1.OA

Add and subtract within 20.

Relate counting to addition and subtraction.

Add and subtract within 20, demonstrating fluency for addition and subtraction within 10. Use strategies such as counting on; making ten; decomposing a number leading to a ten; using the relationship between addition and subtraction; and creating equivalent but easier or known sums.

As you read the book, use a think-aloud method to demonstrate how to use the *I Can* statements:

"On the first page there are 3 red flowers. On each flower there are 6 petals. There is a question for me to answer: How many red flowers were there? I can find the answer to this question by counting the flowers or by reading the words on the page. It says there are 3 red flowers. There is another question: How many pretty petals were there? This question is asking about the total number of petals on the page. But first I have to go back to the number of flowers, which is 3. I know there are 6 petals on each flower because it says there are 6. I have to think about *how the numbers in the problem are related*. There are 6 petals on each flower and there are 3 flowers. So now *I can represent the problem with symbols*. I have to write 6 + 6 + 6 because I am adding three sets of 6. I know that 6 + 6 = 12 and then I add 12 and 6 to get 18. When I am solving a word problem I have to use *the units in the problem* so the answer should be 18 *petals*."

If your students are ready to use the *I Can* statements, lead them into an activity using the tiny black bugs on the page:

- Let's review our answers for the first two questions: How many red flowers were there? (There were 3 flowers.) How many pretty petals were there?" (There were 18 petals.)

- There is one more question on this page: How many tiny black bugs were there in all? This page says that each petal had 2 tiny black bugs. Which number will we use from the answer to the previous question? (We will use 18 because there are 18 petals and the bugs are on the petals.)

- How will we use 18 and 2? What is their relationship? You can use the illustration to help you explain your answer. (There are 18 petals and 2 bugs on each petal. We have to add 18 sets of 2.)

- Let's write the equation for this problem: 2 + 2 + 2 + 2 + 2 + 2 + 2 + 2 + 2 + 2 + 2 + 2 + 2 + 2 + 2 + 2 + 2 + 2. That is a lot of 2s! Is there a way we can write an equation for this problem with fewer numbers? Let's just think about the number of bugs on one flower. (There are 6 petals and each petal has 2 bugs.) What would our equation look like for this part of the problem? (2 + 2 + 2 + 2 + 2 + 2.)

- We know how to count by twos so let's see how many bugs are on this flower. I'll point to each petal on the flower and you count by 2s: 2, 4, 6, 8, 10, 12. So what is our answer? Don't forget to use the units. (There are 12 bugs.)

- Now we know the number of bugs on one flower. There are 3 flowers. How can we make a new equation representing this part of the problem? (12 + 12 + 12.)

- There are many ways we can solve this problem. We can use a calculator. We can also break apart the numbers so we can use numbers that are easier to add. Let's break apart 12 into tens and ones (1 ten and 2 ones, 10 + 2). We have to do that three times: 10 + 2 + 10 + 2 + 10 + 2. We know how to count by tens so how much is 3 tens? (30.) How much is 3 twos? (6.) Now we have 3 tens and 6 ones. What is 30 + 6? (36.) How can we answer the question including the unit? (There are 36 bugs in all.)

Advanced Problem Solvers

How Many Mice? by Michael Garland—Representing and Solving Problems With Addition and Subtraction

The book *How Many Mice?* (2007) can be used by students who are ready to apply all of the components of SMP2. The story follows ten hungry mice that leave their home to gather their meal, encountering an abundance of food as well as several predators on their journey. You can encourage your students to use the illustrations to count the mice and their piece of food and answer the question posed about the mice or the food on each page. Your students can use the events in the story to relate addition and subtraction.

Operations and Algebraic Thinking 1.OA

Represent and solve problems involving addition and subtraction.

Use addition and subtraction within 20 to solve word problems involving situations of adding to, taking from, putting together, taking apart and comparing, with unknowns in all positions, e.g., by using objects, drawings and equations with a symbol for the unknown number to represent the problem.

As you read the book, encourage your students to figure out how many mice and food are on each page by representing the problem on each page with an equation:

■ Let's look at the page where the mice first find some food. Each mouse has a cherry. The question on this page is, "How many cherries can you count?" (10.)

■ We're going to write an equation to represent what is happening on the next page. There are 3 crows who stole 4 of the cherries. The question on this page is, "How many cherries do the mice have now?" We know there were 10 cherries in all on the previous page. There are several numbers to think about but we have to focus on *the units in the problem*. What is this question asking about—mice, crows or cherries? (Cherries.)

■ Yes, cherries. So now we can think about only those numbers. How are *the numbers in the problem related?* What operation is being modeled here? How do you know? (There were 10 cherries and the crows stole 4 of the cherries. That would be subtraction because some cherries are taken from the group.)

■ Now we are ready to *represent the problem with symbols*. I will use the numbers 10 and 4 and the subtraction and equal symbol. I will also use the letter n to represent the number that we will get for our answer. Here is the equation, $10 - 4 = n$.

- What addition problem can you use to help you with this subtraction problem? (4 + 6 = 10). Now we can use *properties of operations* such as relating addition and subtraction to solve the equation, 10 − 4 = n. (It is 6.) What is the answer to the problem on this page using the units? (The mice have 6 cherries.)

- Now I will read the next page so we can write an equation and solve it. We can see the mice picking two red tomatoes. The question on the page is, "How many pieces of food do the mice have?" We have to look back at the answer to the last problem, 6 cherries. In order to make an equation we have to focus on *the units in the problem*. What is this question asking about—mice, food or tomatoes? (Food.)

- Yes, food. So now we can think about only those numbers. How are *the numbers in the problem related?* What operation is being modeled here? How do you know? (There were 6 cherries and then the mice picked 2 tomatoes. That would be addition because there are groups of food being put together.)

- Now we are ready to *represent the problem with symbols*. I want you to work with a partner to create the equation. Make sure you have the numbers, the symbols for addition and the equal symbol and the letter n to represent the number that we will get for our answer. (6 + 2 = n) What is the answer to the problem on this page using the units? (The mice have 8 pieces of food.)

Encourage your students to keep track of the changing number of food items held by the mice prompted by the questions on each page. They should discuss the numbers and how they are related, paying attention to the units in each question so they can dismiss any extraneous numbers. They should also describe the situations to help them decide if addition or subtraction would be the most efficient operation to solve the problem. Then they can create an equation to represent the problem with symbols. Point out opportunities to use the Commutative or Associative Properties of Addition as well as using known addition facts to solve subtraction problems.

Concluding Remarks

Children can begin moving to the abstract stage of problem solving once they have a handle on how to make sense of the numbers in the problem. They have to be exposed to the various addition and subtraction situations specified in Table A.1 in the Glossary of the Common Core State Standards for Mathematics (CCSSI, 2010); only then can they compare the situation in the word problem to the known situations in order to decide whether or not to add or subtract. SMP2 also provides an opportunity for children to become exposed to properties of operations with concrete objects or visuals so they can apply these properties in abstract forms as they move through the grades into more advanced algebra.

3

Construct Arguments

Unpacking the Standard—SMP3: Construct Viable Arguments and Critique the Reasoning of Others

How many times have you witnessed a student obtain the correct answer to a word problem because they just guessed which operation to use with the numbers in the problem? Without even reading the words, children often search for the numbers and try to "do something" with them. As word problems become more complex, involving more than two numbers or requiring multiple steps and operations, children will find it more difficult to solve word problems by just guessing if they do not know how to apply the problem-solving process. If children can learn at an early age to justify their answer, as well as share their reason with others, they can build their repertoire of strategies and get in the habit of crafting an explanation.

Children who are successful at SMP3 construct an explanation to support their answer the minute they are deciding which operation to perform. They have a reason for their decision based on experience examining the various problem-solving situations for addition and subtraction based on Table 1 in the Glossary of the Common Core State Standards for Mathematics (CCSSI, 2010). They can justify their answer with a drawing, a diagram, a concrete model or by acting it out. They know how to listen and ask questions when being exposed to the reasoning of their peers.

In order to fully apply SMP3 when approaching a word problem, children should be able to take ownership of their procedures by using the following *I Can* statements:

- *I can explain my reason for my answer.*
- *I can use objects, drawings, tables and actions to represent the problem.*
- *I can listen and respond to the way others solved the same problem.*

Early Problem Solvers

How Many Seeds in a Pumpkin? by Margaret McNamara—Skip-Counting and Comparing Three-Digit Numbers

The book *How Many Seeds in a Pumpkin?* (2007) can be used to illustrate SMP3 by showing how several students in a class can explain their reason for solving a problem. Early problem solvers should practice listening to the reasons of others before they learn how to respond to the way others solved the same problem. In the story, Mr. Tiffin asks his first grade class to estimate the number of seeds in each of the three pumpkins sitting in front of the class—one small, one medium and one big pumpkin. The students in the story provide reasons for their estimate then count the seeds in each pumpkin by twos, by fives and by tens. Your students can practice skip counting and comparing three-digit numbers with the seeds in the story.

Number and Operations in Base Ten 2.NBT

Understand place value.

Count within 1,000; skip-count by 5s, 10s and 100s.

Compare two three-digit numbers based on meanings of the hundreds, tens and ones digits, using >, = and < symbols to record the results of comparisons.

As you read the book, engage children in a discussion by asking the following questions based on the *I Can* statements:

- In this story, Charlie and his classmates are comparing the size of the three pumpkins that their teacher, Mr. Tiffin, put on his desk one morning. Mr. Tiffin wants them to estimate the number of seeds in each pumpkin. With this book, we can see how students *can explain their reason for their answer.*

- How many seeds did Robert, Elinor and Anna say were in the pumpkin and why did they give that number? (Robert said there were 1 million seeds in the biggest one because the biggest one has the most. Elinor said the medium one has 500 because she always sounded like she knew what she was talking about. Anna said the tiny one has 22 because she liked even numbers.)

- When the children were opening up the pumpkins, what did Robert say as they all looked inside the pumpkins? (He said the big one definitely has the most.) Why do you think he says this? (He was looking at the seeds inside so there must have been a lot in the big one.) How could the students *use objects, drawings, tables or actions to represent the problem?* (They can take the seeds out and count them.) How do they decide to count the seeds? (They lay out all of the seeds on a table. Some students count them by twos, by fives and by tens.)

■ Let's look at the page where one of the groups, The Twos Club, laid out their seeds. How did they arrange their seeds so they could count them easily? (They put them in pairs so there are two seeds together.) Let's count the first row of seeds together by twos, 2, 4 . . . 34. It says there are 170 pairs of seeds so we'll find out later in the story how many seeds there are in the biggest pumpkin.

■ Let's look at the page where one of the groups, The Fives Club, laid out their seeds. How did they arrange their seeds so they could count them easily? (They put them in groups of five seeds together.) Let's count the first row of seeds together by fives, 5, 10 . . . 45. It says there are 63 groups of five seeds and one seed left over so we'll find out later in the story how many seeds there are in the medium pumpkin.

■ Let's look at the page where Charlie, the Tens Club, laid out his seeds. How did he arrange the seeds so he could count them easily? (He put them in two rows of 5 so there are 10 seeds together.) Let's count the first row of seeds together by tens, 10, 20 . . . 70. It says there are 35 groups of ten seeds so we'll find out later in the story how many seeds there are in the smallest pumpkin.

■ It took the Twos Club a long time to count all of the seeds by twos but they ended up counting 340 seeds in the biggest pumpkin. We *can use a table to represent the problem.* Let's keep track of how many hundreds, tens and ones are in each pumpkin (Table 3.1).

TABLE 3.1

Size of pumpkin	Hundreds	Tens	Ones
big	3	4	0
medium			
small			

■ It did not take as long for the Fives Club to count all of the seeds by fives, which was 315, and then they added the one seed left over to count at total of 316 seeds in the medium pumpkin. We *can use a table to represent the problem.* Let's add the hundreds, tens and ones for the medium pumpkin (Table 3.2).

TABLE 3.2

Size of pumpkin	Hundreds	Tens	Ones
big	3	4	0
medium	3	1	6
small			

■ Now we're going to compare the number of seeds in the big pumpkin and in the medium pumpkin. We can use our table (Table 3.2) to compare the hundreds, tens and ones in each number. In order to decide which number is larger, we have to first look at the hundreds. How many hundreds are in the big pumpkin? (3.) How many hundreds are in the medium pumpkin? (3.) Are there more hundreds in the big pumpkin or in the medium pumpkin and how do you know? (They have the same because if we look at the table, there are 3 hundreds in each pumpkin.)

■ So we have to look at the tens next. How many tens are in the big pumpkin? (4.) How many tens are in the medium pumpkin? (1.) So are there more tens in the big pumpkin or in the medium pumpkin and how do you know? (There are more tens in the big pumpkin because 4 tens is more than 1 ten.) We don't have to look at the ones now because we have enough information to decide that 340 is bigger than 316. We can use symbols to record the results: 340 > 316.

■ But Charlie still has to count his seeds. It did not take long at all for him to count 35 groups of 10 gave him a total of 350 seeds in the smallest pumpkin. We *can use a table to represent the problem.* Let's add the hundreds, tens and ones for the smallest pumpkin (Table 3.3).

TABLE 3.3

Size of pumpkin	Hundreds	Tens	Ones
big	3	4	0
medium	3	1	6
small	3	5	0

■ Now we're going to compare the number of seeds in the big pumpkin and in the smallest pumpkin. We can use our table (Table 3.3) to compare the hundreds, tens and ones in each number. In order to decide which number is larger, we have to first look at the hundreds. How many hundreds are in the big pumpkin? (3.) How many hundreds are in the smallest pumpkin? (3.) Are there more hundreds in the big pumpkin or in the small pumpkin and how do you know? (They have the same because if we look at the table, there are 3 hundreds in each pumpkin.)

■ So we have to look at the tens next. How many tens are in the big pumpkin? (4.) How many tens are in the small pumpkin? (5.) So are there more tens in the big pumpkin or in the small pumpkin and how do you know? (There are more tens in the small pumpkin because 5 tens is more than 4 tens.) We don't have to look at the ones now because we have enough information to decide that 350 is bigger than 340. We can use symbols to record the results: 350 > 340.

- What does Mr. Tiffin say about being able to tell how many seeds are in a pumpkin? Think about how you *can listen and respond to the way others solved the same problem*. (The small pumpkin had the most seeds. Mr. Tiffin said you can't tell until you open it up.)

- Mr. Tiffin said there are some clues. What were they? Again, think about how you *can listen and respond to the way others solved the same problem*. (For each line on the outside there is a row of seeds on the inside. The longer the pumpkin grows the more lines it gets and its skin gets darker. That's why the smallest pumpkin had the most seeds. It was dark orange and had the most lines on the outside.)

After reading the book, have your students practice comparing other three-digit numbers based on real world examples such as the number of students in specific grade levels in your school or from articles in the newspaper. They can make a table to record the number of hundreds, tens and one in each three-digit number or create a model with base ten blocks. Then have them record their results with the greater than symbol. They can also practice justifying their reason for choosing the larger number based on the hundreds, tens and ones in each number.

Developing Problem Solvers

Mall Mania by Stuart J. Murphy—Adding Two-Digit Numbers Using Various Strategies

Once your students have improved their listening skills and practiced explaining their answer with an estimation activity, they can learn specific addition strategies. The book *Mall Mania* (2006) is about a group of five children in a chess club who are counting the number of people entering a mall in order to give a prize to the hundredth shopper. Four of the children are stationed at different mall entrances, while the team captain is inside with a walkie-talkie, keeping track of the totals reported each hour. As the children call in their numbers, they use different strategies to add the numbers. Your students can practice using and sharing various strategies for addition.

Number and Operations in Base Ten 1.NBT

Use place value understanding and properties of operations to add and subtract.

Add within 100, including adding a two-digit number and a one-digit number, and adding a two-digit number and a multiple of 10, using concrete models or drawings and strategies based on place value, properties of operations, and/or the relationship between addition and subtraction; relate the strategy to a written method and explain the reasoning used.

As you read the book, stop at various pages in order to allow your students to engage in the following activity, focusing on the *I Can* statements:

- On page 10, the children counted the following number of shoppers at the four mall entrances: 7, 4, 3 and 2. Good problem solvers *can explain their reason for their answer.* Which numbers can you add to make a ten? (7 and 3.) One strategy is to make a ten, then add the remaining numbers. Add up the numbers and be ready to tell the class how you added the numbers together.

- On page 12, we can see that Nicole and Gabby both have 16 as their answer but they used different strategies to add them up. Let's see if you *can listen and respond to the way others solved the same problem.* Listen while I read their explanations. Respond if you used a similar strategy to add the numbers. Did anyone have a different strategy?

- Who *can use objects, drawings, tables and actions to represent the problem?* Please come up to the board and show us how you added the numbers.

- On page 14, the children counted 8, 8, 7 and 7 shoppers. Can you use doubles as a strategy? Add up these numbers and see if you *can explain your reason for your answer.*

- On page 16 we can see that Jonathan and Steven both have 30 as their answer but they used different strategies to add them up. Let's see if you *can listen and respond to the way others solved the same problem.* Listen while I read their explanations. Respond if you used a similar strategy to add the numbers. Did anyone have a different strategy?

- On page 18 the children counted 8, 9, 7 and 8 shoppers. Add up these numbers and see if you *can explain your reason for your answer* by writing a sentence or two explaining how you added the numbers together.

- On page 20 we can see that Steven and Gabby both have 32 as their answer but they used different strategies to add them up. Let's see if you *can listen and respond to the way others solved the same problem.* Listen while I read their explanations. Respond if you used a similar strategy to add the numbers. Did anyone have a different strategy?

- On page 23 the children counted 5, 5, 5 and 6 shoppers. Did anyone make a ten? Add up these numbers and see if you *can explain your reason for your answer* by using some chart paper at your group to write your number sentence and your explanation of how you added the numbers together.

- On page 24 we can see that Jonathan and Nicole both have 21 as their answer but they used different strategies to add them up. Let's see if you *can listen and respond to the way others solved the same problem.* Let's look at the chart paper for each group while someone in the group reads the sentences. Respond if someone in your group used a similar strategy to add the numbers. Did anyone in your group use a different strategy?

Summarize the activity by having students name some of the strategies used in the book to add the numbers. How many different strategies were used? Who used a strategy that was used in the book? Were there strategies used by your students that were not used in the book? How did the explanation and the illustrations of the strategies help everyone understand them?

Advanced Problem Solvers

Spaghetti and Meatballs for All! by Marilyn Burns—Creating Composite Shapes

Students should also practice using SMP3 for geometry and measurement concepts, which can be done with the book, *Spaghetti and Meatballs for All!* (1997). In this book, Mr. and Mrs. Comfort are planning a dinner party for a total of 32 people. They have to rent tables and chairs in order to seat everyone but as soon as guests arrive, the tables start getting pushed together so people can sit together. Children can explore the concept of how to compose new shapes from the composite shape as they use squares to make rectangles and larger squares.

Geometry 1.G

Reason with shapes and their attributes.

Compose two-dimensional shapes (rectangles, squares, trapezoids, triangles, half-circles and quarter-circles) or three-dimensional shapes (cubes, right rectangular prisms, right circular cones and right circular cylinders) to create a composite shape and compose new shapes from the composite shape.

As you read the book, use a think–aloud method to demonstrate how to use the *I Can* statements:

"Let's keep track of the number of people who arrive at the dinner party. So far, there is Mr. and Mrs. Comfort, their daughter, her husband and their 2 children. That is 6 people. When they arrive, they want to push some of the little tables together so they can sit with each other. Mrs. Comfort begins to say, "But there won't be room . . ." So the problem I have to figure out is, will there be enough room for 6 people to sit at the 2 tables after they are pushed together? When I look at the illustration in the book I see there are 4 chairs around each table. That means that 8 people could sit at those 2 tables. I'm going to *use a drawing to represent the problem:*

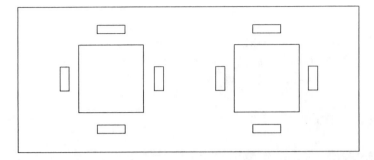

FIGURE 3.1

Now I have to figure out how many people will be able to sit around those two tables after they push them together. I'm going to *use a drawing to represent this part of the problem*:

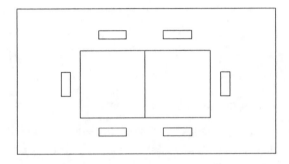

FIGURE 3.2

I can use my drawing to *explain my reason for my answer*. I can see that when they pushed those 2 tables together, they had to remove the 2 chairs on the inside. In my drawing, there are only 6 chairs so that means 6 people can sit around the tables. So instead of 8 chairs there are only 6 chairs that can be used. The answer to this problem is that there will be enough room for 6 people to sit at the 2 tables after they are pushed together."

Continue reading the book and let the students draw a picture to help figure out the next problem:

■ On the next page, 6 more guests arrive and someone suggests pushing two more tables over to the 2 tables that are already pushed together. Mrs. Comfort says, "But that won't work." Let's see what she means by that. How many people are at the dinner party now? There were 6 people and 6 more people arrived. 6 + 6 is 12, so there are 12 people there now.

- Our new problem is to see if there will be enough room for 12 people to sit at the 4 tables after they are pushed together. How can you *use a drawing to represent the problem*?

- Draw the two tables pushed together just like my drawing. How many chairs fit around these tables? (6) Now make another *drawing to represent the problem*. There are 2 more tables pushed together with the first 2 tables:

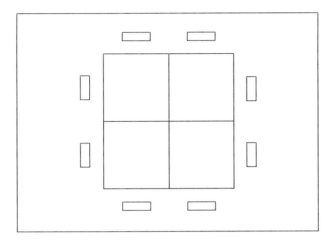

FIGURE 3.3

- Is there enough room for the 12 people to sit at these tables? Turn to your partner and *use your drawing* of the tables and chairs to *explain your reason for your answer*.

Ask for a few students to explain their answer and share their answer. Prompt students to *respond to the way others solved the same problem* by comparing their own explanation and drawing to find similarities and differences. Continue with the other pages in the story in which more guests arrive and more tables are pushed together.

If your students are ready to explore composite shapes, made from combining shapes, read the book again and keep track of the number of tables, number of people who could sit at each table and the new composite shape made when the square tables are pushed together (Table 3.4). Students can use squares cut out of construction paper for the tables so they can rearrange them as in the story.

TABLE 3.4

Number of square tables pushed together	Number of chairs around groups of tables	Representation of table configurations	New shape (composite)
2	6		rectangle
4	8		square
6	10		rectangle
8	12		rectangle
8	16		2 squares
8	18		rectangle
8	20		2 rectangles
8	24		4 rectangles

Concluding Remarks

Children in the early grades are learning to use objects, drawings, diagrams and actions to represent word problems while they are building up their speaking, reading, writing and listening skills. They can practice their speaking skills by explaining their reason for their answers. As their literacy skills increase, children can label their drawings and diagrams and begin to write sentences to explain their reason. Children can practice explaining their representation to their peers and when they are able to respond to each other's explanation, they are demonstrating they are listening carefully.

Create a Model

Unpacking the Standard—SMP4: Model with Mathematics

Most teachers will agree that children should engage in problem solving related to their lives, yet many teachers only use the word problems included in their math curriculum. It is rare to find that every word problem is relevant to every child in your classroom, so care must be taken to individualize or modify word problems at times in order to fit the lives of your learners. This will make it easier for the children in your class to have some background knowledge related to the word problem as well as understand the language and context of the problem in order to work through the problem-solving process.

Children who are successful at SMP 4 can understand how word problems are connected to their daily lives, both in school and out. They can locate the relevant quantities in word problems and are proficient in their knowledge of the situations for the operations so they can represent these quantities. They can utilize the strategy of simplifying a problem when needed in order to make sense of the quantities and their relationships. They can also look back at the original problem in light of their mathematical model to determine if the answer makes sense.

In order to fully apply SMP 4 when approaching a word problem, children should be able to take ownership of their procedures by using the following *I Can* statements:

- *I can solve problems in everyday life.*
- *I can identify important quantities and represent their relationships.*
- *I can simplify a problem.*
- *I can reflect on the results to see if they make sense.*

Early Problem Solvers

The Doorbell Rang by Pat Hutchins—Representing Addition in Various Ways

In the book *The Doorbell Rang* (1986) a mother is baking chocolate chip cookies and offers a plate of the cookies to her two children, Victoria and Sam. Just as the children decide to share the 12 cookies among themselves, the doorbell rings and 2 more children come in. The story continues as more children arrive and Sam and Victoria agree to share the 12 cookies evenly among the group of children. At the end, Grandma rings the doorbell with another batch of cookies. You can use the examples of sharing the cookies to help your students represent addition problems through objects, acting out the situations from the book and creating equations. Provide your students with 12 mini chocolate chip cookies or use circle counters to represent the cookies.

Operations and Algebraic Thinking K.OA

Understand addition as putting together and adding to, and understand subtraction as taking apart and taking from.

Represent addition and subtraction with objects, fingers, mental images, drawings, sounds, acting out situations, verbal explanations, expressions, or equations.

As you read the book, engage children in the following activity based on some of the *I Can* statements:

- "How many of you have eaten cookies or some other type of treat that you had to share with others? It is good manners to offer treats to others rather than to eat in front of someone who does not have the treat. It is also good manners to share the treats so each person has an equal amount. In this book, we'll see 2 children, Victoria and Sam, who show they have very good manners by sharing the plate of cookies with their friends. This is one way we *can solve problems in everyday life.*"

- After reading the first 2 pages, stop and ask, "How do we know the number of cookies that are on the plate? Sam and Victoria say they will get 6 cookies each." Have the students turn to a partner and discuss how they can figure out the total number of cookies on the plate while you pass out a resealable sandwich bag with 12 brown circles representing the cookies. Allow students to tell you how many cookies they think are on the plate, accepting all answers and writing them on the board. Ask students to justify their answers.

- "I know there are 12 cookies because the 2 children each get 6 cookies. *I have identified these important quantities so now we can represent their relationships* with the equation, $6 + 6 = 12$. We know that Sam will get 6 cookies, so this is what the first 6 represents in our equation. We also know that Victoria will get 6 cookies, so this what the other 6 represents in our equation. We know that 6 plus 6 equals 12. So there are 12 cookies on the plate." Draw a picture to show how the 12 cookies are shared among the two children (Figure 4.1).

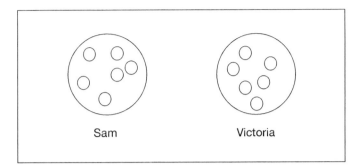

FIGURE 4.1

■ Allow the students to open the resealable sandwich bag of cookie counters and make 2 piles of 6 cookies. Then tell them to count all of the cookies *to see if the results make sense.* They should be able to justify that 6 + 6 equals 12. Continue reading the story and stop after the next set of pages when Tom and Hannah come to the door and Ma says they have to share the cookies.

■ "There were only 2 children sharing the 12 cookies at the beginning of the story but now there are 2 more children at the door. How many children are there now?" (4.) "Use your cookie counters to figure out how many cookies each child will get if they share the 12 cookies equally. You can use a dry erase board to draw 4 plates in order to evenly divide up the 12 cookies."

■ Have a student draw the representation on the board, document camera or Smart Board to demonstrate how to divide up the 12 cookies among the 4 children (Figure 4.2). Have the other students check their physical model to see if they are correct. "Did Sam and Victoria get the same amount of cookies, more cookies or fewer cookies when they had to share them with Tom and Hannah?" (They get fewer cookies.)

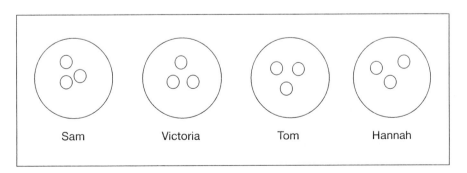

FIGURE 4.2

■ On the next page, your students can check their answer as Sam and Victoria say they will each get three cookies. "Let's see if we *can reflect on the results to see if they make sense.* In this part of the story, the numbers of cookies are the *important quantities and that is what we can represent* with our equation, 3 + 3 + 3 + 3 = 12."

■ The doorbell rings again and 2 more children come in. "There were 4 children and now 2 more children have come to the door but there are still 12 cookies. In this part of the story, the numbers of children are the *important quantities and that is what we can represent* with an equation. Show me an equation to represent the number of children who now have to share the cookies." The students can use their dry erase board to write the equation 4 + 2 = ? Call on a student to explain what the 4 represents (the number of children already in the kitchen) and what the 2 represents (the number of children at the door). "We can count the children in the picture or solve the equation by adding 4 plus 2. There are 6 children now. Use your cookie counters to figure out how many cookies each child will get if they share the 12 cookies equally. You can use the dry erase board to draw 6 plates to evenly divide up the 12 cookies. Do you think Sam and Victoria will get the same amount of cookies, more cookies or fewer cookies now?" (They will get fewer cookies.) "How do you know?" (Every time they share cookies with more people, everyone gets fewer cookies.)

■ Have a student draw the representation on the board, document camera or Smart Board to demonstrate how to divide up the 12 cookies among the 6 children (Figure 4.3). Have the other students check their physical model to see if they are correct:

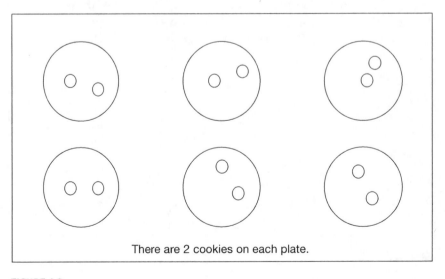

There are 2 cookies on each plate.

FIGURE 4.3

- "Let's see if we *can reflect on the results to see if they make sense*. In this part of the story, the numbers of cookies are the *important quantities and that is what we can represent* with our equation, $2 + 2 + 2 + 2 + 2 + 2 = 12$. As I read the next page, we can check our answer. Yes, we are correct because Sam and Victoria say they will each get 2 cookies."

- The doorbell rings again and 6 more children come in. "There were 6 children and now 6 more children have come to the door but there are still 12 cookies. In this part of the story, the numbers representing the children are the *important quantities and that is what we can represent* with an equation. Show me an equation to represent the number of children who now have to share the cookies." The students can use their dry erase board to write the equation, $6 + 6 = ?$ Call on a student to explain what the first 6 represents (the number of children already in the kitchen) and what the other 6 represents (the number of children at the door). "We can count the children in the picture or solve the equation by adding 6 plus 6. There are 12 children now. Use your cookie counters to figure out how many cookies each child will get if they share the 12 cookies equally. If you need to use the dry erase board you will have to draw 12 plates to evenly divide up the 12 cookies."

- Have a different student draw the representation on the board, document camera or Smart Board to demonstrate how to divide up the 12 cookies among the 12 children (Figure 4.4). Have the other students check their physical model to see if they are correct:

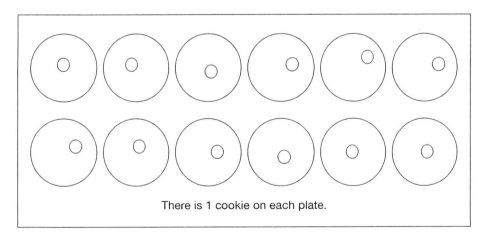

There is 1 cookie on each plate.

FIGURE 4.4

- "Let's see if we *can reflect on the results to see if they make sense*. In this part of the story, the numbers representing the cookies are the *important quantities and that is what we can represent* with our drawing of the 12 plates and 1 cookie on each plate. As I read the next page, we can check our answer. Yes, we are correct because Sam and Victoria say they will each get one cookie. What would happen if more children come to the door?" (They can break the cookies in half to share them. They can hurry and eat the cookies before Ma opens the door. They can ask the children to come back later when Ma makes more cookies.)

The doorbell does ring again but it is Grandma with another tray of cookies. Now students can practice other combinations of cookies to be shared equally among the children. Your students can relate to the real life problem in this story even if their family does not bake chocolate chip cookies by discussing other treats their family cooks, bakes or purchases. You can create word problems related to the story and allow students to use objects, drawings, or equations, reinforcing the concept that as the number of people increase, each person's share gets smaller.

Developing Problem Solvers

Bigger, Better, Best! by Stuart J. Murphy—Using Addition With Rectangular Arrays

Another familiar context to use with your class besides the kitchen is their bedroom. In the book *Bigger, Better, Best!* (2002) three young siblings are constantly arguing about whose belongings are the best. One day their parents announce they will be moving to a new house and will each have their own room. When they go to their new house to select their rooms, they argue over which bedroom window is the biggest. Their mother suggests covering each window with sheets of paper to see which window has the largest area. When they argue about which bedroom is the biggest, their father suggests covering the floor with newspaper to see which room has the larger area. Your students can explore the concept of using addition to find the total number of units in a rectangular array while they are introduced to the concept of area.

Operations and Algebraic Thinking 2.OA

Work with equal groups of objects to gain foundations for multiplication.

Use addition to find the total number of objects arranged in rectangular arrays with up to 5 rows and up to 5 columns; write an equation to express the total as a sum of equal addends.

As you read the book, engage children in the following activity based on the *I Can* statements:

■ "How many of you have your own bedroom? Who has to share a bedroom? In this story, Jill and Jenny are sisters who share a bedroom. Sometimes they argue with their brother Jeff to compare their belongings, like a backpack or book, to see whose are better. Maybe you argue with a brother, sister, cousin, neighbor or friend too. Let's see how their parents help them settle their arguments when they go see their new house."

■ Read the story and stop on page 13 when Jeff and Jenny begin arguing over whose bedroom window is bigger. "What was their mother's suggestion for checking to see which window is bigger?" (She told them to use paper to cover their windows. The window that uses more paper to cover it is the one that has the larger area.) Introduce the term *area*, defined as the number of square units to cover the inside of a shape.

■ "Yes, by using the same size sheets of paper, they can see if it takes more sheets to cover one of the windows. This is one way we *can solve a problem in everyday life.*"

■ Read through page 15 where Jeff completely covers the inside of his window with the sheets of paper. "How many sheets of paper did Jeff use to cover his window? Let's see if we can *identify important quantities* on this page *and represent their relationships.* How many sheets of paper does he have on the window?" (There are 12 sheets.)

■ Continue reading through page 16 where Jenny begins covering her window with sheets of paper. "If we look at page 17, how many sheets of paper can Jenny fit across the length of the window? The *length* is the longer side." (She can fit 6 sheets across.) Introduce the word *array* if your students are not familiar with the term, defined as a set of objects arranged in *rows* and *columns*. "How many rows of paper can she fit into her window? Rows go across the longer side of the window." (She can fit 2 rows across.) "We can make an equation to figure out the number of sheets she will use to cover the inside of the entire window by adding the number of sheets in each row. There are 2 rows and she can fit 6 sheets in each row. So the equation would be 6 + 6 = ? What is the answer?" (12 sheets.)

■ "We can *reflect on the results to see if they make sense.* I will read page 17 so we can check our answer." As you read the page, verify with the students 12 sheets of paper will cover the window. "We can go back to the problem in the story now and decide whose window is bigger by comparing the total number of sheets that covered each window. It took 12 sheets of paper to cover each window. Even though the windows are a different shape, they can have the same area."

■ Continue reading page 18 where the siblings are arguing over whose room is bigger. "I see the father on the next page and he has some newspaper in his hand. What do you think he will tell Jenny and Jeff to do with the newspaper? Think about what they did with the sheets of paper in the window." (He will tell them to cover the floor with the newspaper to see which room is bigger.)

■ Read page 19 and show pages 20 and 21 to your students, "How can Jenny figure out the number of sheets of newspaper that will cover the whole floor when she only has newspaper taped along two of the walls? Sometimes we can *simplify a problem* to help us understand it." Show the class how you can use square inch tiles to start covering a 6 inch by 2 inch rectangular figure:

■ "If this is my bedroom and I start covering the floor with tiles, I can see there are 6 tiles that fit along one of my walls and 2 tiles that fit along the other wall. How can I figure out how many tiles will cover the whole floor when I only have tiles along two of the walls? I have to *identify the important quantities* which are 6 and 2. There are 2 rows of 6 tiles. I can *represent their relationships* by making an equation, 6 + 6 = ? The first 6 represents the first row of tiles and the other 6 represents the second row of tiles. I know 6 plus 6 equals 12 so there will be 12 tiles that will cover the whole floor."

■ "Now we'll go back to the book and look at Jenny's room. There are 6 sheets of newspaper that fit along one of her walls and 5 sheets that fit along the other wall. How can we figure out how many sheets will cover her whole floor? We have to *identify the important quantities* which are 6 and 5. There are 5 rows of 6 sheets of newspaper. We can *represent their relationships* by making an equation, 6 + 6 + 6 + 6 + 6 = ? The first 6 represents the first row of newspaper, the next 6 represents the second row of newspaper, the third 6 represents the third row, the fourth 6 represent the fourth row and the fifth 6 represents the fifth row. If I go back to my equation, **6 + 6** + 6 + 6 + 6, I already know 6 plus 6 equals 12 so there will be 12 sheets in the first two rows. Then I go back to the equation, 6 + 6 + **6 + 6** + 6, and see there is another 12, which is the third and fourth rows. Then there is one more 6 left in the equation, 6 + 6 + 6 + 6 + **6**, which is the fifth row. Now I have to add 12, 12 and 6."

■ Allow time for the students to figure out the answer. They can add 12 and 12 to get 24 then count up 6 more to get 30. They can add all of the numbers by regrouping tens and ones. They can draw a picture or use manipulative tiles to add the sheets of newspaper in their visual or concrete model to get 30. If they have a calculator, explain which buttons to press so they can represent 12 + 12 + 6 = on their screen to get 30.

■ "We can *reflect on the results to see if they make sense*. I will read page 20 so we can check our answer." After you read the page say, "We were right. It would take 30 sheets to cover the floor. Now I'll read another page so we can see how many sheets it will take to cover Jeff's floor."

■ Read page 22 to your students, "How can Jeff figure out the number of sheets of newspaper that will cover the whole floor when he only has newspaper taped along two of the walls? Remember how we can *simplify a problem* to help us understand it." Show the previous 6 inch by 2 inch rectangular figure and prompt the students to use the same procedure to figure out how many sheets of newspaper will cover Jeff's room.

■ "Now we'll go back to the book and look at Jeff's room. There are 6 sheets of newspaper that fit along one of his walls and 4 sheets that fit along the other wall. How can we figure out how many sheets will cover his whole floor? We have to *identify the important quantities* which are 6 and 4. There are 4 rows of 6 sheets of newspaper. We can *represent their relationships* by making an equation, 6 + 6 + 6 + 6 = ? The first 6 represents the first row of newspaper, the next 6 represents the second row of newspaper, the third 6 represents the third row and the fourth 6 represent the fourth row. Before we solve this equation, can you tell if there will be more or fewer sheets that will cover Jeff's room? How *can you reflect on the results to see if they make sense?*" Allow time for your students to turn to a partner and discuss how they can answer your question by comparing the equations (There are 5 rows of 6 sheets in Jenny's room but only 4 rows of 6 sheets in Jeff's room so it will take more sheets to cover Jenny's room.).

- "If I go back to my equation, **6 + 6** + 6 + 6, I already know 6 plus 6 equals 12 so there will be 12 sheets in the first two rows. Then I go back to the equation, 6 + 6 + **6 + 6**, and see there is another 12, which is the third and fourth rows. Now I have to add 12 and 12."

- Allow time for the students to figure out the answer. They can add 12 and 12 to get 24 or they may recall that 24 is the answer from this step from the previous problem. They can add the numbers by adding the tens and ones. They can draw a picture or use manipulative tiles to add the sheets of newspaper in their visual or concrete model to get 24. If they have a calculator, explain which buttons to press so they can represent 12 + 12 = on their screen to get 24.

- "We can *reflect on the results to see if they make sense.* I will read page 23 so we can check our answer." After you read the page say, "We were right. It would take 24 sheets to cover the floor."

- Read page 24 and show page 25 so your students can see that there is another area in front of the closet that is part of the area of Jeff's room. There are 2 rows of 3 sheets. Have your students represent this with a visual model by drawing the rows of sheets or with a physical model by using manipulative tiles to represent the 2 rows of 3 sheets. Then have your students write an equation to represent the 2 rows of 3 sheets.

- "I can see from your drawings and your tile models that you have a row of 3 and another row of 3. Use your dry erase boards to write your equation and solve it this time too." (3 + 3 = 6). We know 24 sheets would cover the larger area of his bedroom and 6 sheets will cover the smaller area. How many sheets will cover the whole floor in his room?" Remind your students of the problem regarding the sheets of paper in Jenny's room and how the first step was adding 12 and 12 to get 24 then adding the last 6 to get 30.

- "We can *reflect on the results to see if they make sense.* I will read page 25 so we can check our answer." After you read the page say, "We were right. It would take 30 sheets to cover the floor. So if the area of Jeff's room is 30 and the area of Jenny's room is 30, is one of their rooms bigger?" (The area of both rooms is the same so one room is not bigger than the other.)

This book can also be used to introduce or reinforce the concept of area. Students should practice with a consistent unit of measure, such as square inches, when comparing the area of two or more figures or objects. They can also use inch graph paper to highlight equal rows of square units represented in the story. As they highlight each row of squares, they can write one of the addends in their equation.

Advanced Problem Solvers

Alexander, Who Used to be Rich Last Sunday by Judith Viorst—Solving Word Problems With Money

All children can relate to the real life situation of using money. In the book *Alexander, Who Used to be Rich Last Sunday* (1978) the main character, Alexander, is complaining that his two brothers have money when he only has bus tokens. He is thinking about the previous Sunday when his grandparents visited their home and gave him and his brothers each one dollar. Throughout the story he recounts how he spent all of the money, little by little, until he had nothing left but bus tokens (you may have to discuss what bus tokens are if students are not familiar with them.) The students should know the coin names and values before they use this book for problem solving. They should each have a set of real coins, with at least 13 pennies, 7 nickels, 7 dimes and 3 quarters. The students will also use play dollar bills as well.

Measurement and Data 2.MD

Work with time and money.

Solve word problems involving dollar bills, quarters, dimes, nickels and pennies, using $ and ¢ symbols appropriately.

As you read the book, engage children in a discussion by asking the following questions based on the *I Can* statements:

- "When is the last time someone gave you money as a gift? Did you save it or spend it? We're going to read a book about a boy named Alexander to find out what he does with the money his grandparents give to him. You are going to use the real coins you brought to school for our money activities. We are going to keep track of the money amounts in the story to see how we *can solve problems in everyday life*"

- We know that good problem solvers *can identify important quantities and represent their relationships*. Today our quantities involve coin amounts and coin values. We will use the real coins and play dollars to keep track of the different coin amounts and values in the story.

- Read the first page then display the following word problem: Anthony has $2, 3 quarters, 1 dime, 7 nickels and 18 pennies in his room. How much money does Anthony have?

- Allow the students to use the play dollar bills and real coins to put out the number of each according to the problem. Do the same on the document camera or Smart Board while you read the problem aloud again. "Now that each one of us has the correct amount of each dollar bill and coin, we can find out the amount of money Anthony has on the first page of the story. First we have to *identify the important quantities*. What is the value of a quarter? (25¢) The problem states Anthony has 3 quarters. How can we *represent the relationship* between the value of the quarter and the number of quarters in his room? (We can write that 3 quarters are worth 75¢). What is the value of a dime? (10¢). The problem states Anthony has 1 dime. How can we *represent the relationship* between the value of the dime and the number of dimes in his room? (We can write that 1 dime is worth 10 cents). What is the value of a nickel? (5¢). The problem states Anthony has 7 nickels. How can we *represent the relationship* between the value of the nickel and the number of nickels in his room? (We can write that 7 nickels are worth 35¢) What is the value of a penny? (1¢.) The problem states Anthony has 8 pennies. How can we *represent the relationship* between the value of the penny and the number of pennies in his room? (We can write that 8 pennies are worth 8¢.) Let's look at the values that we wrote. How much money is that in all? (If your students can add with regrouping, they can list all of the amounts vertically and add the ones, then the tens, to get $1.28.) The problem also states that Anthony had $2. So how much money does he have in his room?" (Anthony has $3.28.)

- Read the second page then display the following word problem: Nicholas has $1, 2 quarters, 5 dimes, 5 nickels and 13 pennies. How much money does Nicholas have?

- Allow the students to use the play dollar bills and real coins to put out the number of each according to the problem. Do the same on the document camera or Smart Board while you read the problem aloud again. "Now that each one of us has the correct amount of each dollar bill and coin, how can we find out the amount of money Nicholas has on the second page of the story?" Let's fill out a table (Table 4.1) to keep track of the type of bills and coins, the values of each and the total value:

TABLE 4.1

Type of money	Value of each	Total number of each	Total value
Dollar	$1	1 dollar bill	$1
Quarter	25¢	2 quarters	50¢
Dime	10¢	5 dimes	50¢
Nickel	5¢	5 nickels	25¢
Penny	1¢	13 pennies	13¢

- Allow students to discuss how they can add up the total value of coins in the last column of the table and put in those values. "In which column did we *identify the important quantities* from the word problem? (In the second column we wrote the value of the dollar and each coin. In the third column we wrote the number of each bill and coin.) How did we *represent the relationship* between the values of the bill and the coins and the number of each bill and coin?" (In the last column we wrote the total value of the bill and coins.)

- "How much money is that in all? If we look at the table, we can make another dollar with the 50¢ from the quarters and the 50¢ from the dimes. Then by adding the 25¢ and the 13¢, we get 38¢. So how much money does he have in his room?" (Anthony has $2.38.)

- "We *can reflect on the results to see if they make sense.* Let's look at the third column in our table to see if it makes sense that he has $2.38. There is only 1 one dollar bill, but we made another dollar with these 2 quarters and these 5 dimes. It makes sense that we have $2. Now we can check the nickels and pennies to see if they add up to 38¢. We can count the nickels by fives then keep counting by ones when we get to the pennies. 5, 10, 15, 20, 25, 26, 27, 28, 29, 30, 31, 32, 33, 34, 35, 36, 37, 38."

- "Now that we have practiced adding up coin amounts and dollar amounts, we are going to keep track of Alexander's spending. His grandfather gives him and his brothers each $1. Since Alexander spends it a little at a time, we are going to first exchange one dollar bill for the following coins: 7 dimes, 4 nickels and 10 pennies. Let's count these coins to be sure they equal one dollar." The students can choose their preferred method for counting coins: counting first by tens, then by fives then by ones; counting the like coins and recording the amounts (70, 20 and 10) then adding those together.

- Once the students have verified they all have the equivalent of one dollar with their coins, continue reading the book. Stop at the page where Alexander buys gum at Pearson's Drug Store. "The last line on this page states Good-bye fifteen cents. What does that mean? (He spent fifteen cents so it is gone.) Use your dimes and nickels to remove 15¢ from your pile of Alexander's money. How can we use the least amount of coins to make 15¢? (We can use one dime and one nickel.) Remove one of your dimes and one of your nickels because he spent 15¢ out of his dollar so far. Put those coins on the other side of your desk. How much money does he have left right now?" Allow students to count the remaining coins and write down all answers on the board.

- "We *can reflect on the results to see if they make sense.* If Alexander had one dollar, then he spent 15¢, the amount he has left and the amount he spent should add up to one dollar. Let's check the answers to see which one works. (Alexander has 85¢ left.) Tell me how you found your answer. "Allow students to explain how they knew he had 85¢ left.

- Read the next two pages. Stop at the page where he has to pay his mom. "The last line on this page states Good-bye another 15¢. We are going to use the same 2 coins to represent the 15¢. Remove one of your dimes and one of your nickels and put them on the other side of your desk. How much money does he have left now?" Allow students to count the remaining coins and write down all answers on the board.

■ "We *can reflect on the results to see if they make sense.* How much did Alexander spend so far and how do you know? (30¢ because 15¢ and 15¢ is 30¢.) If Alexander had one dollar, then he spent 30¢, the amount he has left and the amount he spent should add up to one dollar. Let's check the answers to see which one works. (Alexander has 70¢ left.) Tell me how you found your answer." Allow students to explain how they knew he had 70¢ left.

Continue reading the book and stopping on the pages where he spent money in order to discuss how to remove the amount using the fewest coins, counting the remaining amount of money and reflecting on the results to see if they make sense by adding up the total amount spent and their answers to see if they equal one dollar. As your students continue to share their strategies for counting money, emphasize the most efficient strategy and see if all of the students can count the coins that way.

Concluding Remarks

There are many ways for children to model the mathematics in the word problems they encounter. They can use a physical model, visual model, number model, create a table, or use written or oral procedures. By allowing students to choose how they want to represent the word problem, they can use the one that is best for them at their current stage of problem solving. They may also choose a model based on the type of word problem. When teachers use word problems with a familiar context, students can spend less time trying to understand the language in the word problem and more time making sense of the quantities and their relationships.

5

Use Mathematical Tools

Unpacking the Standard—SMP 5: Use Appropriate Tools Strategically

Children love to use calculators, rulers, tape measures and other mathematical tools, but do they know how and when to use them in their everyday lives? Teachers explicitly instruct their students how to use a ruler to measure an object. Then the students practice measuring objects on a workbook page and in the classroom. The same is often true with calculators; teachers show their students which buttons to press to calculate an answer and then the students practice pushing buttons to find the answer to calculations on a workbook page. Teachers need to spend some additional time explicitly instructing their students how to use tools such as these when solving word problems.

Children who are successful at SMP 5 know which types of mathematical tools are available to them as well as the functions of each tool. They have practiced using the tools to solve everyday problems such as measuring space in the classroom in order to move a bookshelf or adding large numbers on a calculator to determine how many students will be attending a field trip. They can determine whether or not mental math or a calculator would be more efficient to solve a problem or if technology would be useful to solve a problem. They are also familiar with how tools, such as graphs, maps or concrete models, can be used both within and outside of mathematical problem solving.

In order to fully apply SMP 5 when approaching a word problem, children should be able to take ownership of their procedures by using the following *I Can* statements:

■ *I can learn how to use different mathematical tools.*

■ *I can choose the right tools to solve a problem.*

Early Problem Solvers

Earth Day-Hooray! by Stuart J. Murphy—Using Place Value to Add and Subtract

In the book *Earth Day-Hooray* (2004) a group of children are cleaning up a neighborhood park for an Earth Day activity. They realize that if they take the cans they find to the recycling center, they can earn money to buy flowers to plant in the park. At school, their club adviser helps them keep track of their cans in groups of tens, hundreds and eventually thousands in order to achieve their goal of 5,000 cans to turn in for money. Your students can use models, drawings and strategies based on place value to solve problems from the book. You can have your students use drawings, a calculator, and base ten blocks and mat as tools to solve the problems in the book.

Number and Operations in Base Ten 2.NBT

Use place value understanding and properties of operations to add and subtract.

Add and subtract within 1,000, using concrete models or drawings and strategies based on place value, properties of operations and/or the relationship between addition and subtraction; relate the strategy to a written method. Understand that in adding or subtracting three digit numbers, one adds or subtracts hundreds and hundreds, tens and tens, ones and ones; and sometimes it is necessary to compose or decompose tens or hundreds.

As you read the book, engage children in discussion and activities by asking the following questions based on the *I Can* statements:

■ How do scientists use tools to help them? (Scientists use tools to measure things, to dig in the ground, to explore space, to mix chemicals, to fix things, to take things apart, etc.) Good mathematical problem solvers can also *learn how to use different mathematical tools*. These tools can be things like calculators, rulers, pencil and paper, graphs, a hundreds chart, a number line, drawings or objects. We are going to use some of these types of tools as we solve problems from this book.

■ The children in this story are part of the Maple Street School Save-the-Planet Club and they are cleaning up a park near their school. What kind of garbage do you see the children picking up on pages 4 and 5? (They are picking up wrappers, scraps of paper, newspapers, bottles, cans and cups.) On page 6, Ryan realizes that the park would look better with some flowers at the entrance. How could they get money to buy flowers? (They could ask their parents, use their allowance, do a fundraiser at school, etc.) Have you heard of a recycling center? Ryan talks about taking the aluminum cans they are collecting to a recycling center so they can get money for the cans. These centers buy cans from people and then use them to make new cans. What do you think Ryan and his friends would do with the money they get from turning in the cans? (They could buy the flowers to plant in the park.)

- They tell their club adviser about it and she said they should collect 5,000 cans! How could they keep track of that many cans? I am going to give you a minute to talk to a friend about how the children can keep track of all of the cans they collect so they will know when they reach 5,000. (They can count them, put them in stacks, write down how many they collect, etc.)

- Thanks for sharing all of the ways you discussed with your friend. I am going to show you page 10 so you can see how they sorted the cans and put them into groups of ten. Why do you think it will be easier for them to count the cans if they are in groups of ten? (Because they can count by tens faster than by ones.) It is easier and faster to count by tens and then they can keep track of the groups of hundreds as they collect more cans. How many groups of ten will they need in order to have a group of 100 cans? (They would need 10 groups of ten.)

- As I read the book, I am going to tell you the groups of cans they are collecting, in ones, tens, hundreds and eventually in thousands but you will have to keep track of the amounts. Let's think about how we *can choose the right tools to solve a problem*. What tools can you use to keep track of the amounts of cans? (We can draw pictures, make a graph, use a calculator, etc.)

- Let's try drawing pictures as a tool to help us solve some of these word problems. I am not going to show you the picture on page 11 until after you have drawn a picture so you can check your work. I will write a word problem that goes with this page so you can look at it as you draw your picture:

  The children collected three big bags of 100, five small bags of ten and 9 single cans. How many cans did they collect so far?

- Think about how you could draw bags and label them. Just draw a rectangle for a can so it won't take long. When you are finished, show your drawing to a friend sitting near you and figure out how many cans the children collected. Write the answer on the top of your paper and circle it.

- How did you know that there were 359 cans? (We added the hundreds, which was 300. Then we added the tens, which was 50. Then we added 300 + 50 + 9, which is 359.) You remembered to group all of the hundreds together, then group the tens together and then add them up with the ones. How did using your drawing as a tool help you figure out the problem? (I could see how many bags of 100 they collected and how many bags of ten and then I could draw 9 cans.)

■ As I continue to read the story, we can see on page 17 that the children had to start all over because the trash collector took away their cans. But now they are starting over again with 56 cans. On pages 17, 18 and 19 how are they able to get more cans since they already cleaned up the cans from the park? (They are putting up signs around the school, going to each classroom and picking up cans around the neighborhood.)

■ I am going to give you another word problem to solve and this time you can use another tool. I will give you base ten blocks for you and your partner to use as models for the hundreds, tens and ones. Here is the problem that has two parts to the solution:

> The children brought the cans they collected to school to count at recess. There were six bags of 100, three bags of 10 and five single cans. How many cans did they count at recess? How many cans did they collect so far if they combine those cans with the 56 cans they already collected?

■ Now use your base ten blocks as a tool to find out how many cans they counted at recess. Don't forget to group the hundreds together, the tens together and the ones together before you add them up to find your total. How many cans did they count at recess? (They counted 635 cans.) How did using the base ten blocks as tools help you figure out the problem? (We could group the hundreds blocks together to see there were 600, then we grouped the tens blocks together to see there were 30 and then we had 5 ones. We added 600 + 30 + 5 together to get 635.)

■ Now we have to solve the second part of the problem. If they counted 635 cans at recess and they already had 56 cans, how many cans have they collected altogether? You might need to compose a ten when you are adding the tens and ones. Use more base ten blocks to add to the 635 that you already have on your base ten mat.

■ How many cans did they collect altogether? (They collected 691 cans.) How did using the base ten blocks and mat as tools help you figure out this part of the problem? (We added 6 ones to the 5 ones that we already had on our mat. That made 11 ones so we exchanged 10 of those for a tens block. Then we put that tens block and 5 more tens blocks on the tens section with the 3 tens that we already had there. That made 9 tens. We didn't add any more hundreds so that made 691.)

■ You are doing a great job keeping track of the hundreds, tens and ones. I think you're ready to try the thousands now! We are going to do one more word problem but we will *learn how to use different mathematical tools.* We are going to use a calculator and paper and pencil so we can see the numbers and make sure we write them correctly. Here is the word problem that has two parts:

> The children collected cans from students and teachers all week. They collected one bag of 1,000, four bags of 100, eight bags of 10 and 3 single cans. How many cans did they collect that week? How

many cans will they have if they add that number to the 691 cans they already collected?

■ Let's do the first part together. How many bags of 1,000 did they collect? (One bag.) So we are going to write 1,000 on our paper. I will write it on the Smart Board so you can see if you wrote it correctly. How many bags of 100 did they collect? (Four bags.) So now we have to add 100 four times on our calculator and write down our answer. What should we press on our calculator to get 100 in the screen? (We press the 1 key then the 0 key then the 0 key again.) Okay, once you have 100, you will have to press the + key and then put in another 100 and then press the + key again. Do you have 200 in your screen? If so, you did it correctly. If not, clear it and try it again or look for a friend sitting near you who did it correctly and you can do it together. Now put in 100 again. If you press the + key you should see 300 in your screen. Then put in 100 again and press the = key and you should have 400 as your answer.

■ Write 400 below 1,000 on your paper, but line up the hundreds. We will do the tens now. There are 8 bags of ten. That will take a long time to put in the calculator if we do it the same way we did the hundreds. I will show you a shortcut using the x key on your calculator. There are 8 bags of 10 so we can use multiplication as a shortcut. Press the 8 key then press the x key then press the 10 key. Now press the = key. You should see 80 in your screen. 8 x 10 will give us the same answer as adding 8 groups of 10.

■ Write the 80 below the 400 but line up the tens. There were also 3 single cans so we have to write 3 below the 80 but line it up with the ones:

    1,000
      400
        80
         3

■ How many cans did they collect that week? (They collected 1,483 cans.) How did using the calculator, paper and pencil as tools help you figure out the first part of the problem? (We could figure out how many hundreds and tens using the calculator and then write the other numbers on the paper.)

■ We have one more part of the problem to solve. How many cans have they collected altogether? Let's use our calculator to solve this problem but I still want you to write the equation on the paper so you can see the numbers you will be putting into the calculator. What is the equation for this part of the problem? (It is 1,483 + 691 =?) Let's all write that equation on our paper. Now put that into your calculator. Be sure to press all of the number keys correctly for 1,483 and then the + key. You don't have to press anything for the comma that is between the 1 and the 4. Now put in 691 and press the = key. You should see 2,174 on your screen.

You can continue reading the book and challenge your students to solve the final problem in the book, which is on pages 28 and 29, where the children finish collecting the cans and get their grand total to be able to turn in the cans to the recycling center for money to purchase flowers. Your students can work on this last two-part problem with a partner in class or for homework. They can also choose the tool or tools they would like to use to help them figure out the answers. Ask the students why they chose the tool and how it helped them solve the problem.

Developing Problem Solvers

Mummy Math by Cindy Neuschwander—Recognizing Attributes of 3D Objects

The book *Mummy Math* (2005) can be used to introduce or reinforce the names and attributes of solid geometric shapes. The characters Matt and Bibi are twins who take a trip to Egypt with their parents who are scientists. Once they arrive at one of the Egyptian pyramids, Matt and Bibi get lost inside but realize they can use clues based on the geometric solids to help them find their way out. When using this book, have the following geometric solids on display: cone, cylinder, cube, sphere, square-base pyramid, tetrahedron (triangle-base pyramid), rectangular prism and triangular prism with each solid labeled with the name. It is suggested to have a few other sets of these solids in order to pass around to groups of students.

Geometry 2.G

Reason with shapes and their attributes.

Recognize and draw shapes having specified attributes, such as a given number of angles or a given number of equal faces.

As you read the book, engage children in discussion and activities by asking the following questions based on the *I Can* statements:

■ When Matt and Bibi arrive in Egypt, they see the great pyramids that were built thousands of years ago to bury kings and other important people. Let's look at the eight geometric solids I have here. Which one is the same shape as the pyramid on pages 2 and 3? (The square-base pyramid.) Yes, it is the pyramid with the square at the bottom. There are two types of pyramids. One has a triangle as the bottom, or *base*, and one has a square as the base.

■ On pages 7 and 8, it looks like Matt and Bibi are lost but they see a clue on the wall to help them find their way out. It says there are many faces to guide them. What do you think of when you hear the word *face*? (A person's face.) In geometry, a face is the flat side of a geometric solid. Some solids have no faces, like the sphere. Some have 1 face, like a cone. Some have 2 faces, like the cylinder, and some have more than 2 faces. You *can learn how to use different mathematical tools* when you are solving problems if you know the names of the objects you can use as tools.

■ I am going to have you sit in groups while I continue to read the story so I can put a set of these solids in the middle of your group. Spread them out and then each take the solid closest to you. Look for the flat parts, which are the faces. How many faces are on the triangle-base pyramid? (4 faces.) How many faces are on the square-base pyramid? (5 faces.) How many faces are on the triangular prism? (5 faces.) How many faces are on the rectangular prism? (6 faces.) How many faces are on the cube? (6 faces.)

■ Let's make a table of the solids and the faces so we can refer to it while we solve the problems in the story (Table 5.1). I will also ask you about the shape of each face so we can fill in that column:

TABLE 5.1

Geometric solid	Number of faces	Shape of each face
Cone	1	Circle
Cylinder	2	Circles
Tetrahedron (triangle-base pyramid)	4	Triangles
Square-base pyramid	5	1 square 4 triangles
Triangular prism	5	2 triangles 3 rectangles
Rectangular prism	6	6 rectangles or 2 squares and 4 rectangles
Cube	6	6 squares

■ Matt and Bibi will have to use their knowledge of the number of faces on each of the geometric solids to guide their way out. The geometric solids are the tools that will help to solve the problems in the story. Let's look at one of the first clues on page 12, "A single face shows the way." Let's look at our table to see which geometric solid shown on this page has only 1 face, the cone or the rectangular prism? (The cone has only 1 face.) Yes, the cone is the clue so that means they have to go up because the flat part of the cone is facing up. This shows how you *can choose the right tools to solve a problem.*

■ Which geometric solid do they see when they climb up to the ledge? (They see a sphere.) According to our table, how many faces does a sphere have? (A sphere has no faces.) Right, there are no faces to follow but their only choice is to crawl through a tunnel into a large room where they find the next clue, "Look for six identical faces." What does *identical* mean? (Identical means exactly the same.) So on pages 16 and 17, Matt and Bibi have to find a geometric solid that has all faces exactly the same size and shape.

By looking at our table, which solid do you think they are looking for and why? (They are looking for a cube because it's the only solid on those pages that has 6 identical faces.) Yes, the cube has 6 faces, which are each a square. Again, you are showing that you *can choose the right tools to solve a problem.*

- Now they are in another room with three towers made up of solid shapes. On page 20 they say they are to "enter under the five faces." We can use our table, but remember there are two solids stacked on top of each other so we have to think about the clue and the tools used to solve the problem. You can use the solids in your group to create the towers and look at the table if you need it. The first tower is a cylinder with a cone on top. How many faces would they walk under? (They would walk under 2 faces, the top of the cylinder and the bottom of the cone.) The second tower is a triangular prism with a triangle-base pyramid on top. How many faces would they walk under? (They would walk under 5 faces, the top of the prism and the 4 faces of the pyramid.) The third tower is a rectangular prism with a square-base pyramid on top. How many faces would they walk under? (They would walk under 6 faces, the top of the prism and the 5 faces of the pyramid.) So which tower should they choose? (They should choose to walk into the second tower.)

- It looks like Matt and Bibi did choose the second tower and found the burial chamber and the pharaoh, or king, who is buried there. What is the last type of tool they use to help them find their way out of the tower? (They use a map on the coffin lid to find their way out.)

As a follow-up activity, create word problems based on the number and shape of the faces on the solids. The answer to each problem should be one of the geometric solids. As a challenge, have students create their own word problems based on the solids for homework. You may want to send home a copy of the table for a reference. An example of a word problem would be:

Matt and Bibi are looking for a solid that has only 2 faces. Each face is a circle. Which solid are they looking for? (A cylinder.)

Advanced Problem Solvers

Measuring Penny by Loreen Leedy—Measuring in Standard and Nonstandard Units

The book *Measuring Penny* (1997) can be used to introduce, explore or compare the use of standard and nonstandard units of measure. Lisa's teacher gives the class a homework project that requires the students to measure something at home using standard and nonstandard units. Lisa chooses to measure her dog Penny, using standard units such as inches, centimeters, pounds and nonstandard units such as dog biscuits and cotton swabs. For this story, have the following objects available: ruler, yardstick, cotton swabs, measuring spoons, measuring cups and a balance scale.

Measurement and Data 2.MD

Measure and estimate lengths in standard units.

Measure the length of an object by selecting and using appropriate tools such as rulers, yardsticks, meter sticks and measuring tapes. Measure the length of an object twice, using length units of different lengths for the two measurements; describe how the two measurements relate to the size of the unit chosen.

As you read the book, engage children in discussion and activities by asking the following questions based on the *I Can* statements:

- Let's look at the examples of standard and nonstandard units that Lisa's teacher, Mr. Jayson, listed on the chalkboard. What do you think is the difference between *standard* and *nonstandard* units? (Standard units are the kind that we use in school or at home like inches, gallons or minutes. Every inch is the same, every gallon is the same, every minute is the same. Nonstandard units are not the same. They can be anything you find like paper clips or pencils. Not every paper clip is the same. Not every pencil is the same.)

- Lisa is going to have to *learn how to use different mathematical tools* to measure her dog Penny so she uses both standard and nonstandard units. Mr. Jayson also requires the students to make at least one comparison so Lisa will have to use words such as, "taller than" or "heavier than," to compare Penny to other dogs. What is the first tool she uses to measure Penny and the two other dogs at the park? (A ruler.) What unit is she using? (Inches.) How does she use the ruler to measure the dogs? (She measures their noses and compares the length in inches.)

- Which nonstandard tool does Lisa use to measure Penny and three other dogs at the park? (Dog biscuits.) When using nonstandard units, the tool is often used as the unit as well. So she is using dog biscuits as the tool and the unit. Who has the shortest tail and what is the length? (Penny has the shortest tail. It is 1 dog biscuit long.)

- Lisa wants to measure each dog's ear using cotton swabs. Are those a standard or nonstandard unit? (Nonstandard unit.) How would she use the cotton swabs to measure the dog's ears? (She would see how many cotton swabs would be the same length as one of the dog's ears.) Next, she wants to measure the width of each dog's paw print. How does she use the ruler to measure their paw prints? (She uses the centimeter side of her ruler. Each dog makes a paw print in the sand. She measures each paw print with the ruler and writes the width in centimeters.)

- When we are measuring objects, people or animals, we have to *choose the right tools to solve a problem*. If we are measuring something very small, like a dog's paw print, would we use a ruler or a yardstick and why? (A ruler because it is smaller than the yardstick.) Yes, it would be easier to use the ruler, even though the yardstick can also be used to measure an object in inches. So we have to think about the size of the object and the size of the tool and choose the right tool. Lisa wants to see how tall the dogs are compared to Penny. What tool is she using on this page and why do you think she chose that tool? (She is using a yardstick because it is tall like the bigger dogs.) What units could she use to record the height of the dogs? (She could use inches, feet or yards). She can also decide which unit to use based on the height of the dog. The Dachshund is 12 inches tall. Why did she write *1 foot* next to it? (Because 12 inches is the same as 1 foot.) In her report, she wrote Penny's height using three different units, but they are all the same, or *equivalent*. Penny is 18 inches tall, which is equivalent to 1½ feet and ½ yard.

- On the next page, Lisa wants to see how high each dog can jump. What tool could she use? (She could make a mark and then measure it with a yardstick or a ruler.) She decides to use herself to measure, with units such as shoulder-high or waist-high. Is this standard or nonstandard? (nonstandard) What other nonstandard tools could she have used? (She could stack books or use the playground equipment.)

- Have you ever played on a seesaw at a park? It is like a balance scale. If the two people weigh the same, the seesaw will be balanced. But if one person is heavier, that side will go down toward the ground. Let's try this out with the balance scale. If I put the same object on each side, like a marker, the pans on the scale will be balanced because the markers weigh the same. But if I replace one marker with a pair of scissors, the pan with the scissors goes down because the scissors are heavier than the marker.

- Remember when Mr. Jayson said that the students have to include a comparison in their report? How does Lisa use the seesaw as a tool to compare the weight of the dogs? (She put Penny on one side and then other dogs on the other side to see if each dog is heavier, lighter or the same as Penny.) Lisa doesn't use any numbers to compare the weight of the dogs, just the comparison with another dog. But on the next page she weighs Penny on a scale like the type a person uses to weigh himself. The unit she uses is pounds.

- On the next page I see some measuring spoons and cups, and containers with the words pint, quart and gallon. We don't use these tools to measure length, height or weight. These tools are used to measure *volume*, which is the amount of space something fills like milk in a container.

- Another way in which Lisa measures Penny is with time. What are some ways to measure time? (Using a stopwatch, a clock or with a calendar.) She makes a schedule to keep track of what Penny does at different times in the day. She also records the amount of time it takes to do tasks that involve taking care of Penny such as feeding and walking her. What units can be used to measure time? (seconds, minutes, hours, days, weeks, months, years).

■ Lisa also measures Penny using temperature. What are some words you think of when you hear *temperature*? (Thermometer, weather, Fahrenheit, Celsius, degrees, hot, cold, freezing.) What tool would you use to measure the temperature? (A thermometer, look it up on the Internet).

■ The last way she uses measurement is with money. What units do we use when we are measuring the cost of something? (Dollars and cents.) What tool can help us if we have a lot of numbers to add? (A calculator.)

After you finished reading the book to the class, the students can do some of the following activities to practice using mathematical tools strategically:

– Students can fill in a table with all of the tools used in the book for standard measurement with a separate column for the units and one for the type of measurement best suited by the tool. (A ruler is used for length, width or height.)

– Have the students list the nonstandard units used in the book. Then they can look around the classroom and write down at least five more objects that could be used for nonstandard units of measure.

– Create stations for students to work in small groups to measure objects with provided standard and nonstandard units.

– Assign a homework project similar to the project in the book so students can practice measuring an object, person or animal using several different types of standard and nonstandard tools and units.

– Choose addition or subtraction problems (single and double-digit) that can be solved with paper and pencil or with a calculator. Have half the class use paper and pencil and the other half use a calculator to see which tools are more efficient for each type of problem.

– Have students create three comparison statements using objects in the classroom and the balance scale.

– Students can create a table with conversions using inches, feet and yards (they should have rulers and yardsticks available to help them make their table). Then give them word problems to solve using the table, a ruler or a yardstick.

Concluding Remarks

Children should be taught to use mathematical tools to help them solve word problems when appropriate. But first, they should become familiar with the types of tools available, which type of measurement is best suited by the tool, how to use the tool, which tool is most efficient, and which units are recorded in their answer. The best types of practice activities allow students to choose the objects to be measured, like Lisa in the story who chose to measure her dog Penny.

Attend to Precision

Unpacking the Standard—SMP 6: Attend to Precision

Most teachers have experienced the frustration of asking students how they arrived at their answer only to hear, "I just knew it." It may seem odd that students do not know how they solved a word problem even if their answer is correct, but more often than not, students simply cannot verbalize their solution process. In order to provide an explanation, a student must possess strong content knowledge, an understanding of the goal of the word problem and why the answer makes sense.

Children who are successful at SMP 6 know how to interpret the symbols and the mathematical language in a word problem. They know the definition of mathematical terms used in their grade level as well as how to use the terms to explain their solution process. They can articulate the problem-solving strategy they applied to the word problem; how they used drawings, diagrams, or concrete objects to arrive at a solution; and which unit will be used to represent their answer.

In order to fully apply SMP 6 when approaching a word problem, children should be able to take ownership of their procedures by using the following *I Can* statements:

- *I can define the meaning of mathematical symbols.*
- *I can correctly label my diagram, drawings, graphs and units in the answer.*
- *I can explain how I solved a problem using mathematical terms.*

Early Problem Solvers

If You Were a Quadrilateral by Molly Blaisdell—Identifying Quadrilaterals

Children can learn the definition of a quadrilateral and related mathematical terms in the book *If You Were a Quadrilateral* (2010). The characters in this book are animals that are using quadrilaterals, which appear in the environment such as a rectangular tennis court, squares on a checkerboard and exercise mats shaped as various parallelograms. On the left side of each two-page spread, the sentence begins with "If you were a quadrilateral. . . ." The terms used in the book are mathematically correct but are written in a way that a child could understand, with pictures and examples used for support. You should have large examples of the following quadrilaterals to display as well as smaller versions for each child or pair of children to share: rectangle, square, trapezoid and rhombus that is not a square.

Geometry 2.G

Reason with shapes and their attributes.

Recognize and draw shapes having specified attributes, such as a given number of angles or a given number of equal faces. Identify triangles, quadrilaterals, pentagons, hexagons and cubes.

As you read the book, engage children in discussion and activities by asking the following questions based on the *I Can* statements:

- In the same way we learn what good readers do; we can learn what good math problem solvers do. One thing good math problem solvers do is *use mathematical symbols and terms correctly*. In this book we will learn about quadrilaterals and there will be many terms we will put up on our Math Word Wall. It says on page 3 that a quadrilateral is "a flat, closed figure with four straight sides." Which figures on pages 4 and 5 are quadrilaterals? (The game board and the squares on the game board; the sheets of paper and the rectangles on the sheets of paper; the patch of grass where the animals are playing ball; the spaces on the hopscotch game.) Yes, these are all figures that are flat, like a piece of paper, and they have 4 straight sides.

- On page 9 there is a wall with a sign that says, Polygon Art Contest, with pictures of shapes under the sign. By looking at the shapes, what do you think would be a definition for a polygon? (It is a shape. It has 4 sides. The sides are straight. It can be any color. Some sides can be longer or they can all be the same size. Some sides can slant.) The definition for a polygon is that it's a closed figure with at least 3 straight sides. The polygons on this page all have 4 sides so they are called quadrilaterals. Triangles and shapes with 5 or 6 sides could be polygons too. But polygons have straight sides so a circle is not a polygon. A closed figure means there can't be any openings. The sides all have to be touching another side.

- On page 10 we can see another important thing to remember about quadrilaterals, which is that they have 4 angles. If we look at the picture we see animals playing baseball on the baseball diamond. The corners where they are on a base or home plate is the angle. It is where 2 of the sides of a polygon are touching.

- Now we can review some shapes you should already know. I will read part of page 12 to see if you can *use the mathematical terms correctly* to figure out which quadrilateral from our display (rectangle, square, rhombus that is not a square) is the one on this page. Here is the first clue, "Each pair of opposite sides would be the same length." There are many terms in this clue that we have to make sure we understand. Good math problem solvers learn math terms or if they don't know it, they find out what it means. Let's see who can come up to the quadrilaterals and show me opposite sides.

- Now let's see who can come up to the quadrilaterals and show me opposite sides that are the same length. Which quadrilateral could be the one on the page? (It could be a rectangle, rhombus or square). Yes, it could be one of those three quadrilaterals. Now I'll read the second clue, "All the angles are the same size." Could it be this rhombus (pointing to the non-rectangle rhombus)? (No because two angles are bigger than the other two.) You're right, not all of the angles are the same size. Could it be the rectangle or the square? (Yes, it could be either one.) I'll show you the page and you can see that there is a tennis court which is a rectangle.

- Here is another problem for you to solve. I will read part of page 14 to see if you can *use the mathematical terms correctly* to figure out which quadrilateral from our display (rectangle, square, rhombus that is not a square, trapezoid) is the one on this page. Here is the first clue, "All sides would be the same length."

- Now let's see who can come up to the quadrilaterals and show me sides that are all the same length. Which quadrilateral could be the one on the page? (It could be rhombus or square). Yes, it could be one of those two quadrilaterals. Now I'll read the second clue, "All the angles are the same size." Could it be this rhombus (pointing to the non-rectangle rhombus)? (No because two angles are bigger than the other two.) You're right, not all of the angles are the same size. I'll show you the page and you can see that there is a checkerboard, which is a square and is made up of smaller squares.

- Let's read about a rhombus, which people sometimes call a diamond, but a diamond is not a correct mathematical term. I want you to use the set of quadrilaterals that I gave you for this next set of problems. I'll read the description and I want you to hold up any quadrilateral that matches the description. Here is the first one, "The sides are all the same length." (They could hold up a square or rhombus.) You could have either the square or the other quadrilateral that has the pointier corners, which is called a rhombus, not a diamond. A square is also a rhombus but it has a special name because all 4 angles are the same size. It also states that the opposite angles are the same size so look at both of these shapes and see how the opposite angles of the rhombus are the same—two are bigger and two are smaller.

- Here is another problem we're going to do with your set of quadrilaterals. Hold up any quadrilateral that matches the description. "Each pair of opposite sides would be the same length and the opposite sides never touch or cross each other." Go ahead and look at your shapes while I read it again if you are still trying to find a match. (They could hold up a square, rectangle or rhombus.) You could have a square, rectangle or rhombus. There is another mathematical term that you're going to learn today, which is called parallel. The opposite sides of these three quadrilaterals are parallel because they never touch or cross each other. These are a special type of quadrilateral called parallelograms.

■ Let's look at the other quadrilateral that doesn't match this description. One pair of opposite sides, the top and bottom sides, called the bases, never touches or crosses the other. But the other pair of opposite sides, the right and left sides, called the legs, would touch if we made the sides longer so they cross each other. Only two of the sides are parallel so we can't call the trapezoid a parallelogram.

After you have finished reading the book, your students can label each quadrilateral and describe each shape. They can refer to the word wall or sit with a partner using the book. Provide graph paper and a straight edge for students to draw each quadrilateral, labeling the sides and angles. Create a center or homework activity with several of each type of quadrilateral (using various colors and sizes) for students to sort into categories such as parallelograms or rhombi. Include other shapes such as circles, triangles and ovals so students can expand their categories into polygons and quadrilaterals.

Developing Problem Solvers

Lemonade for Sale by Stuart J. Murphy—Representing and Interpreting Data

The book *Lemonade for Sale* (1998) can be used to introduce or reinforce the concept of representing and interpreting data as well as labeling a graph. The children in the story want to have a lemonade stand to earn money to repair their clubhouse. They create a bar graph to keep track of the number of cups of lemonade they sell each day for a week. As the week progresses, they fill in the bar graph and compare each day to the previous day, providing opportunities for problem solving involving the information presented in the bar graph. You can create the bar graph in the story with your students by using a large piece of graph paper or recreating it to display on a document camera.

Measurement and Data 2.MD

Represent and interpret data.

Draw a picture graph and a bar graph (with single-unit scale) to represent a data set with up to four categories. Solve simple put together, take apart and compare problems using information presented in a bar graph.

As you read the book, engage children in discussion and activities by asking the following questions based on the *I Can* statements:

■ We can see the children in the story want to sell lemonade in order to earn money to fix up their clubhouse. Danny suggests keeping track of their sales. What is Sheri's idea for keeping track of the sales? (She will make a bar graph and show the number of cups they sell each day.) Let's look at the bar graph on page 7. What information is on the graph? (There are numbers on the left side and days of the week on the bottom.)

■ We have to be sure we *can use mathematical symbols and terms correctly* when we solve word problems so we are going to look closer at the numbers and words on Sheri's graph. Which numbers are on the graph and where did she write the numbers on the graph? (The numbers are from 0 to 90 but they are only the tens. They are next to the lines that go across the graph.) The numbers on the left side of the graph are the *scale*. You can see there isn't room to fit all of the numbers from 0 to 90 so she only put the tens on the scale. She put them next to each *horizontal line* starting at the bottom, which would be 0.

■ Now let's look at the words on the bottom of the scale. What words did she write and where did she write the words on the graph? (She wrote Mon., Tues., Wed., Thurs. and Fri. for the days of the week. She wrote them under every other column of squares on the graph). She left space between the words so it will be easier to see the different columns once they are colored in.

■ What do the numbers represent? (They represent the number of cups sold each day.) What will she color in if they sell 10 cups on Monday? (She will color in one square above the Mon.) What if they don't sell exactly 10? What if they sell 8? (She will color in the square but not all the way to the 10.)

■ I am going to make the same graph as the one Sheri made so we can use it to solve problems about the cups of lemonade sold. I will put the numbers 0 through 90 on the left side of the graph, making sure I am writing only the tens on the scale and that each number is next to a horizontal line. Next I will write the abbreviations for the days of the week under the squares, leaving one column between each day.

■ There is one more thing we have to do before we can continue reading the story. We also have to be sure we *can correctly label diagrams, drawings and units in the answer.* Sheri did not label the graph in the book but I am going to label ours. If someone looked at our graph and did not know we were reading *Lemonade for Sale*, they would not know what the numbers and days of the week represent. What can we write on the left side, next to the numbers, that can let people know what these numbers represent? (They are the number of cups of lemonade they sold.) I will write Number of Cups.

■ Now we have to label the bottom of the graph too, where the days of the week are written. What do these days represent? (They are the days the kids are selling their lemonade.) I will write Days Selling Lemonade. The last thing we have to add is a title. We have to let people know this graph is being used to keep track of the number of cups of lemonade sold. How about, Cups of Lemonade Sold. The title of a graph should be short and is usually not a whole sentence so we don't put a period at the end but we do capitalize the important words.

■ Now we can start filling in the graph as we read the story and then solve some problems with the graph. On the first day of their lemonade stand Sheri said they sold 30 cups. What should we color in to show they sold 30 cups of lemonade? Be sure you can explain why. (Color in 3 boxes above the Mon. because each box is 10. Three boxes will show 30 cups sold.)

■ On the second day of their lemonade stand Sheri said they sold 40 cups. What should we color in to show they sold 40 cups of lemonade? Be sure you can explain why. (Colour in 4 boxes above the Mon. because each box is 10. Four boxes will show 40 cups sold.) Now we can use the graph to solve some problems. I'm going to show you a word problem and I want you to use our graph to solve it. Be sure to label your units and be sure you *can explain how you solved the problem using the mathematical terms.*

> The kids sold 30 cups of lemonade on Monday. They sold 40 cups of lemonade on Tuesday. Did they sell more lemonade on Monday or Tuesday? How many more?

■ I want you to write down your answer to the first question on a dry erase board and hold it up. I see that most of you wrote Tuesday. First, we have to know how to read the bar graph. How do you know how many cups were sold on Monday and Tuesday? Explain your answer using terms such as scale, horizontal line and bar graph. (I know they sold 30 cups on Monday because there are 3 boxes colored in above the Mon. and the colored boxes go up to the horizontal line next to the number 30 on the scale. I know they sold 40 cups on Tuesday because Sheri is drawing a line from the 40 on the scale across the horizontal line to the top of the colored boxes above Tues.) Why do you think they sold more cups on Tuesday? (Forty is a bigger number than 30 so they sold more on Tuesday.) Yes, they sold more on Tuesday. The second question is, how many more? (They sold 10 more cups.) There is more than one way to figure out this answer. What are some ways you know the answer is 10? (You could count up from 30 to 40; you could see that there is one more box colored in on Tuesday and each box is for 10 cups; you could make a subtraction problem, which is $40 - 30$.)

■ I heard someone say that it could be a subtraction problem which would be written, $40 - 30 = ?$ This is called a comparison situation where the difference is unknown (see Appendix 1). We could also solve it with an addition situation where the addend is unknown, which would be written, $30 + ? = 40$. If you said you could solve the problem by counting up from 30 to 40, then you were using the addition situation with the addend unknown.

- I'm going to give you another word problem that you can solve by looking at the graph on page 14 again. Show me your answer and write an addition or subtraction situation that would match the problem:

    The kids sold 30 cups of lemonade on Monday and 40 cups of lemonade on Tuesday. How many cups of lemonade did they sell over both days?

- Show me on your dry erase board how you solved this problem. (30 + 40 = 70) How did you know that you would add the numbers? (We had the first amount of lemonade, which was 30, and another amount, which was 40, and then we had to put the two amounts together). Yes, this is an example of a *put together* addition situation with the total unknown.

    Continue reading the story, making up word problems to go with each page that your students can solve as a class, with a partner or individually based on their skill level and familiarity with the addition and subtraction situations. They can then create their own bar graph with similar information provided (number of candy bars sold in a week), using a checklist of the components needed to correctly label their graph:

    - Write the numbers on the left side of the graph for your scale.
    - Label the left side of the graph.
    - Write words at the bottom of the graph.
    - Label the bottom of the graph.
    - Write a short title for the graph at the top.

Advanced Problem Solvers

Measuring Penny by Loreen Leedy—Relating Addition and Subtraction to Length

In addition to using *Measuring Penny* (1997) to teach your students how to use standard and nonstandard units while applying SMP 5, use appropriate tools strategically, you can revisit the story or use it for the first time for SMP 6. When Lisa is measuring Penny and other dogs in her neighborhood, she uses drawings and diagrams to keep track of her results. Lisa also uses various mathematical symbols and terms and labels her units, which vary throughout the book. You and your students can use examples in the book to create word problems relating addition and subtraction to length.

Measurement and Data 2.MD

Relate addition and subtraction to length.

Use addition and subtraction within 100 to solve word problems involving lengths that are given in the same units, e.g., by using drawings (such as drawings of rulers) and equations with a symbol for the unknown number to represent the problem.

As you read the book, engage children in discussion and activities by asking the following questions based on the *I Can* statements:

■ Now that you are advanced problem solvers, you should be able to *use mathematical symbols and terms correctly* in order to solve word problems and *explain how you solved a problem using mathematical terms*. In this book, Lisa is given a homework assignment, which requires her to measure something in as many ways as she can, record her results by including the number and unit, making at least one comparison and using both standard and non-standard units. Her teacher, Mr. Jayson, provides examples of units they can use but he tells them to be creative. Lisa will have to *correctly label her diagram, drawings, graphs and units in the answer*.

■ When Lisa gets home, she gets the idea to measure her dog, Penny. She is already thinking about comparing Penny to at least three other dogs in her neighborhood and draws a picture of Penny and the other dogs, labeling each type of dog. Let's look at page 7 where she is measuring the length of each dog's nose with a ruler. I'm going to give you a word problem to solve using this picture:

> Penny is measuring the length of the nose of some of the dogs using her ruler. The sheepdog's nose is 4 inches, Penny's nose is 1 inch and the pug's nose is half an inch. How many inches longer is the sheepdog's nose than Penny's nose?

■ In order to solve this, let's think about whether we would add or sub tract in order to find the difference in length of the dogs' noses. I am going to list the common addition and subtraction situations and I want you to think about which one fits this word problem. The situations are *add to, take from, put together, take apart* and *compare*. Which one fits this word problem? (We are going to *compare* the lengths of their noses.) In this type of comparison, we subtract the smaller length from the larger length. I want you to use your dry erase boards to write an equation with a symbol for the unknown number. You can use L for length.

■ Let me see your equations. The correct equation should be $4 - 1 = L$. I know you can subtract these numbers with mental math so go ahead and erase the L and put in the length. (3.) Did you label your unit? (3 inches.)

■ Are dog biscuits a standard or nonstandard unit and how do you know? (They are a nonstandard unit because all dog biscuits aren't the same size you wouldn't always know how long something is in dog biscuits.) Even though the biscuits are nonstandard units, we can still use addition and subtraction to compare length. Now I'm going to give you another word problem for the picture on page 8:

> Lisa is measuring the length of the dogs' tails using dog biscuits. The greyhound's tail is 10 dog biscuits long. The terrier's tail is 4 dog biscuits long. How many more dog biscuits would Penny need to add to the 4

biscuits on the terrier's tail so that it would equal the 10 biscuits on the greyhound's tail?

- In order to solve this, let's look at the list of the common addition and subtraction situations so we can think about which one fits this word problem. The situations are *add to, take from, put together, take apart* and *compare*. Which one fits this word problem? (We are going to *compare* the number of dog biscuits that are on the dogs' tails.) This is also a comparison situation but it is different than the example with the length of the dogs' noses. They are both using length and they are both comparing the length, but how is the goal of this problem different than the other one? (This problem wants us to figure out how many biscuits would equal 10 and 10 is the big number. In the other problem we put the big number at the beginning of the equation and we subtracted 1.)

- I want you to use your dry erase boards to write an equation with a symbol for the unknown number. You can use B for biscuits. Remember what your friend said about how the equation should equal 10 in this problem.

- Let me see your equations. The correct equation should be $4 + B = 10$. I know you can figure this out with mental math so go ahead and erase the B and put in the number. (6.) Did you label your unit? (6 dog biscuits.)

- For the first problem we wrote $4 - 1 = L$ and that is called a difference unknown type of situation because we didn't know the difference between the two amounts. In the second problem we wrote $4 + B = 10$ and that is called an addend unknown because we didn't know the other number to add that would equal 10.

- Let's look at another way Lisa measured Penny on pages 25 and 26. On these pages, Lisa is showing how she can measure the temperature in degrees Fahrenheit that Penny prefers when she goes on her walk. Lisa drew pictures of Penny and labeled the temperature in degrees at the bottom of each picture. I am going to give you a word problem about temperature:

    Penny likes to go for a long walk when it is 60°F outside. When it is 80°F outside, Penny thinks it is too hot for a long walk. How many degrees hotter is 80°F than 60°F?

- In order to solve this, let's look at the list of the common addition and subtraction situations again so we can think about which one fits this word problem. The situations are *add to, take from, put together, take apart* and *compare*. Which one fits this word problem? (We are going to *compare* 80°F with 60°F.) This is another comparison situation. Is it like the problem with the length of the dogs' noses where we wrote $4 - 1 = L$ or is it like the problem with the length of the dogs' tails where we wrote $4 + B = 10$? Is there a way we can solve this by addition or by subtraction, depending upon how we write the equation? I want you to turn to your partner and talk about how you could create an equation using the addend unknown and an equation using the difference unknown?

- I want you to use your dry erase boards to write an equation you and your partner discussed, using a symbol for the unknown number. You can use T for temperature. It is okay if you and your friend only created one equation so far.

- The equation for the difference unknown is the most common equation to use for this type of problem. The equation should be $80 - 60 = T$. If you wrote an equation using the addend unknown, it would be $60 + T = 80$. In fact, when you are doing your subtraction for $80 - 60$, you might be thinking about what the missing addend would be. For example, I know that it is sometimes faster to think about addition problems when trying to solve subtraction problems. For example, I know that $10 - 4 = 6$ because I know $4 + 6 = 10$.

- It doesn't matter which equation you and your friend are using to solve this word problem because the answer would be the same. What is $80 - 60$? (20.) What number would I add to 60 so that it equals 80? (20.)

You and your students can use many of the other pages to create word problems that can be solved by drawing a picture or diagram or writing an equation. Students should be able to see multiple ways of solving the same problem from the story. Practice using different units in the story such as centimeters, feet and cups.

Concluding Remarks

Students should be as diligent about learning the meaning of their math terms as they are about learning to spell words for a spelling test or studying vocabulary words in science. Teachers have to provide ways for their students to read, define, record, label and practice using mathematical terms. Then when their students are ready, they will be able to understand how the terms are being used in word problems and will be able to use the terms in the explanation of their problem-solving process.

7

Look for Structure

Unpacking the Standard—SMP 7: Look For and Make Use of Structure

Once young children learn how to recognize patterns they begin to point out patterns in their environment including in their classroom and in their home. However these are typically visual patterns such as a sequence of red and blue cubes or repeating shapes on a curtain. Teachers have to guide their students toward recognition of patterns and structure in other aspects of mathematics: digits end in 5 then 0 when skip counting by fives; the number of sides of a polygon increase when the number of angles increase. As children begin to recognize these patterns and structure they will begin to use them as shortcuts to become more efficient at problem solving.

Children who are successful at SMP 7 search for patterns that can be used to reach a solution. For example, they know how to apply counting principles of even and odd numbers based on patterns. They also look for elements that can be grouped together such as addends that are doubles or shapes that have four sides. They know the importance of learning mathematical terms so they can be precise as they explain their problem-solving process for example, saying the word *quadrilateral* when speaking of polygons with four sides.

In order to fully apply SMP 7 when approaching a word problem, children should be able to take ownership of their procedures by using the following *I Can* statements:

- *I can find a pattern in a problem.*
- *I can look for ways to make groups in a problem.*

Early Problem Solvers

The Button Box by Margarette S. Reid—Identifying Attributes for Sorting

In the book *The Button Box* (1990) a young boy plays with the buttons that are in a box at his grandmother's house. He sorts them in different ways such as by size, color and type. Your students can learn what it means to sort objects as well as how to use various types of attributes to classify objects into categories. Once they can identify an attribute and sort objects into categories, they can practice counting the number of objects in each category. Have a collection of buttons similar to the type of buttons in the book. You can also have buttons (real buttons or circles cut out to look like buttons) in a little resealable sandwich bag for each student, that are all the same size but in three different colors. Make sure each student has the same number of buttons of each color so you can assess them. In order to make it easier for your students to sort the buttons into three groups, put three squares of paper in the resealable sandwich bag too.

Measurement and Data K.MD

Classify objects and count the number of objects in each category.

Classify objects into given categories; count the numbers of objects in each category and sort the categories by count.

As you read the book, engage children in discussion and activities by asking the following questions based on the *I Can* statements:

■ Let's look at the picture of the box of buttons that the boy in the story just opened up at his grandmother's house. There are many buttons and they are all different. I have a collection of buttons that I brought from home. As I read the story, we will stop and look at the buttons in the book and the buttons in my collection.

■ The story states that he likes to *sort* the buttons because there are so many different kinds of buttons. He will sort them into different groups, called *categories*. As I read I want you to listen for the first two different categories of buttons. (There are sparkly buttons and buttons covered with cloth.)

■ Let's look at the buttons in my collection. I am going to put some of them on the document camera so we can all see them. Are any of my buttons sparkly? (Yes.) How many are sparkly? (3.) Are any of my buttons covered with cloth? (Yes.) How many are covered with cloth? (2.)

■ Now I am going to read the next two pages. Listen again for some categories for sorting our buttons. (There are metal buttons and leather buttons.) I am going to put another set of my buttons on the document camera. Do I have any metal buttons? (Yes.) How many metal buttons are in the book? You count them while I touch each one in the illustration. (1, 2, 3, 4.) How many leather buttons are in the book? You count them while I touch each one in the illustration. (1, 2, 3, 4.) There are 4 metal buttons and there are 4 leather buttons. There is the same number of metal buttons as leather buttons.

■ Listen for the next category of buttons. What do these buttons look like? (They have designs on them. Some are silver and some are yellow or gold. Some have a flag or a star.) These are the shiny buttons that come from uniforms. I am going to put some more of my buttons on the document camera. Do I have any shiny buttons that look like the buttons in the book? (Yes.) How many? (4.)

■ Let's look at the page where the boy arranged the buttons in rows by color and by size. I am going to give you each a little resealable sandwich bag of circles that we are going to use as buttons. You are going to use them to solve some word problems. Here is the first word problem:

I have a bag of buttons. They are all the same size and shape. How can I sort them?

■ Pour out your buttons onto your desk. What do you notice about the buttons? (They are all the same size but are different colors.) To help us solve this we *can look for ways to make groups in a problem*. So how can we sort them into groups or categories? (We can sort them by color.) We can make a category for each color. I also have a little resealable sandwich bag of buttons that are the same as yours so I will put them on the document camera. Let's do one together. What is one of the colors of buttons in your resealable sandwich bag? (Red.) Okay, let's sort the red buttons first. We have a category called Red Buttons and I am going to create a table so we can keep track of the categories (Table 7.1.) What other colors do you have in your resealable sandwich bag? (Blue.) We also have blue buttons so we also have a category called Blue Buttons that I will put on our table. Are there any other colors in your resealable sandwich bag? (Green.) Yes, there are green buttons too so our last category is called Green Buttons.

TABLE 7.1

Category	Picture of buttons	Number of buttons
Red Buttons		
Blue Buttons		
Green Buttons		

■ Now you can sort your buttons into the three different categories. I put three squares of paper in your resealable sandwich bag so take them out and use one for each color. Sort the red buttons by putting all of the red buttons on one of the pieces of paper. How many red buttons do you have? (5.)

- Now sort the blue buttons by putting all of the blue buttons on another piece of paper. How many blue buttons do you have? (4.) Now sort the green buttons by putting them on the last piece of paper. How many green buttons do you have? (6.)

- Now let's draw the amount of buttons in each category on our table (Table 7.2). How many buttons will I draw for the category Red buttons? (5.) Why should I draw 5 there? (Because there are 5 red buttons in our resealable sandwich bags.) How many buttons will I draw for the category Blue buttons? (4.) Why should I draw 4 there? (Because there are 4 blue buttons in our resealable sandwich bags.) How many buttons will I draw for the category Green buttons? (6.) Why should draw 6 there? (Because there are 6 green buttons in our resealable sandwich bags.)

TABLE 7.2

Category	Picture of buttons	Number of buttons
Red Buttons	☺ ☺ ☺ ☺ ☺	
Blue Buttons	☺ ☺ ☺ ☺	
Green Buttons	☺ ☺ ☺ ☺ ☺ ☺	

- Now we can use our table to help us solve another word problem. Here is the next word problem:

  I have a bag of buttons. I sorted the buttons into three categories. Are there more red, blue or green buttons?

- How can we fill in the rest of our table to help us solve this problem (Table 7.3)? (We can count up the number of buttons in each category and write it in the last column.) What should I write for the number of red buttons? (5.) What should I write for the number of blue buttons? (4.) What should I write for the number of green buttons? (6.)

TABLE 7.3

Category	Picture of buttons	Number of buttons
Red Buttons	☺ ☺ ☺ ☺ ☺	5
Blue Buttons	☺ ☺ ☺ ☺	4
Green Buttons	☺ ☺ ☺ ☺ ☺ ☺	6

- We still have to solve the word problem, which asks if there are more red, blue or green buttons. What is your answer and how do you know? (There are more green buttons because 6 is the biggest number; because the row of green buttons is the longest; because there are only 5 red buttons and 4 blue buttons, etc.

Continue reading the story and talk about other categories of buttons that are in the book. You can give students resealable sandwich bags of buttons that vary in size but are all the same color, so they have to figure out the new category, practicing with only one attribute at a time such as size, color or type of button. Create word problems based on the new category and encourage students to *look for ways to make groups in a problem.*

Developing Problem Solvers

Patterns in Peru by Cindy Neuschwander—Describing Relative Positions

The book *Patterns in Peru* (2007) can be used to teach children how to find patterns in a problem as well as to describe positions of objects. In the story, twins Matt and Bibi are in Peru with their parents who are scientists. They are in an Incan museum when they hear about the Lost City of Quwi (**coo**-ee) and set out to find it. Matt realizes he has the tunic of the secret messenger of the Lost City in his back pack, which is decorated with patterns he and Bibi use to guide their way.

Geometry K.G

Identify and describe shapes (squares, circles, triangles, rectangles, hexagons, cubes, cones, cylinders and spheres).

Describe objects in the environment using names of shapes and describe the relative positions of these objects using terms such as *above, below, beside, in front of, behind* and *next to.*

As you read the book, engage children in discussion and activities by asking the following questions based on the *I Can* statements:

- In this story, twins Matt and Bibi are in Peru, South America with their parents who are scientists. They hear their parents talking to Professor Herrera about a tunic, which is like a shirt, which is over 500 years old. Professor Herrera thinks the tunic belonged to a secret messenger of a lost city.
- Matt and Bibi want to try to find the lost city so they take off on two guanacos (gua-**na**-cos), which are like llamas. Matt gets cold and looks in

his pack for something to put on. He realizes he has the ancient tunic, which their mother said has many patterns on it.

■ Bibi says that the tunic could be the messenger's map. We will see how they *can find a pattern to help them solve a problem.* How can they use the tunic to find the Lost City? (They can see if it shows them where to go. The patterns might be clues.)

■ Now we see patterns on the tunic that have dots in lines slanting up and down. I'm going to put some circle counters on the document camera and I want you to give me directions on how to arrange the counters so they look like the circles on the tunic. I have a list of terms you can use to describe the position of objects (Figure 7.1.)

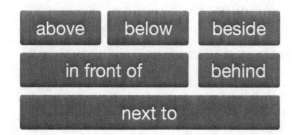

FIGURE 7.1

■ Who would like to go first? (Put 2 circles next to each other.) Okay, I will put 2 of the circle counters next to each other. What else? (Put 2 more circle next to each other but put them below the first 2 circles.) That's great! I like how you used the terms *next to* and *below* to describe the position of the circles. What is the next part of the pattern? (Put 2 circles next to each other and put them beside the other 4 circles.) You're right! And you used the term *beside* so I knew where to put the 2 circles.

■ Matt and Bibi find out that this pattern of 4 circles, then 2 circles shows them that first they go up the mountains on 4 feet, which is when they are riding their guanaco, and then they go down the mountains on their own 2 feet because they are very steep. So they used the pattern of 2 circles, 4 circles, 2 circles, 4 circles, to help them solve their first problem.

■ Now Matt and Bibi are at the edge of a cliff. How can they figure out how to get over the river? (They can look at the tunic to see if there are clues to crossing the river.) Yes, Bibi finds a pattern on the tunic that shows them where to walk on the rope bridge.

■ I have some Popsicle sticks and toothpicks to model the next part of the problem. Let's look at the lines on the tunic. I want you to give me directions on how to arrange the sticks so they look like the lines and Xs on the tunic. Remember to use the list of terms to help you describe the position of objects (Figure 7.1).

■ Who would like to go first? (Put 12 Popsicle sticks *next to* each other.) Okay, I will do that. What should I do with the toothpicks? (Take 2 toothpicks and cross them to make an X. Put those toothpicks *in front of* the third Popsicle stick so it looks like it is crossed out.) Let's see . . . yes, there is an X on the third line on the tunic so I will put the toothpicks on every third Popsicle stick.

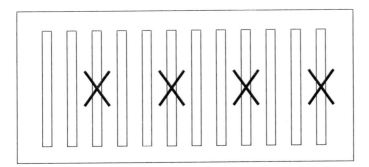

FIGURE 7.2

■ How *can they find a pattern in the problem?* (They knew that they had to step over the ropes that were crossed out when they went over the rope bridge.) Now that they've crossed the river, they find a large carving on a wall but they don't know how to get inside. What can they do to find how to get past the wall? (They can look at the tunic to see if there is another pattern that can be a clue to get inside the wall.)

■ Matt and Bibi can use the positional pattern on the tunic to figure out how they should turn the figure on the carving to get inside. I am going to show you the illustration in the book where Matt is climbing the wall to get past the carving. How can you describe Matt, the dog and the carving in the illustration using the list of terms to help you describe the position of objects (Table 7.4)? (Matt is *in front of* the carving on the wall. The dog is *below* the carving.) Why do you think Matt and Bibi are trying to open the carving? Be sure to use one of the positional terms in your answer. (They think the lost city is *behind* the carving.)

■ Now Matt and Bibi find themselves *in front of* another wall but this one has carvings of animals on it. How *can they find a pattern in the problem?* (They can look at the tunic to see if there are clues to get inside the wall.) Yes, Bibi finds a part of the tunic that tells them how to figure out which animal carvings to touch to get *behind* the wall. But Matt reminds her that some of the tunic was torn off. What does Bibi tell him about patterns? (Patterns are predictable so if they can figure out the first part, they can figure out the rest of the pattern.)

■ We can make a t–chart like the one Bibi makes to see how the numbers get larger for each of the nine panels on the wall. I am going to give you a highlighter and a Hundreds Chart so you *can find a pattern in the problem* and tell me which numbers to write on the t-chart. Let's look at the part of the tunic with the groups of animals. There are 3 foxes, 6 llamas and 9 parrots. On your Hundreds chart, color in the 3, the 6 and the 9 while I write them in the t-chart.

■ Now color in the 12, 15 and 18. Can you predict which numbers you will color in next? How many numbers have you colored in? (6.) Yes, you have colored in six numbers so you have to color in three more numbers. Let's see if we can predict which numbers should be colored in next by looking at the numbers we have already colored in. What is the pattern? (We color every three numbers). Yes, you should be coloring in every three numbers. So which numbers are next? (21, 24, 27.) Go ahead and color in the 21, 24 and 27 in the pattern on your Hundreds Chart while I add them to the t-chart (Table 7.4.) For homework you can color in the rest of the Hundreds Chart with the pattern!

TABLE 7.4

Panel	Number of animals
1st	3
2nd	6
3rd	9
4th	12
5th	15
6th	18
7th	21
8th	24
9th	27

Finish the story and review all of the ways Matt and Bibi used patterns from the tunic to help them solve their problems they faced as they looked for the lost city. You can have your students practice using the positional terms with other objects such as pattern blocks. Give students a basket of pattern blocks and a file folder to put between partners. One student creates a pattern or design with the blocks then uses the positional terms to describe where they placed the blocks to see if their partner can produce the same pattern or design.

Advanced Problem Solvers

The Greedy Triangle by Marilyn Burns—Exploring Attributes of Shapes

Teachers can use the book *The Greedy Triangle* (1994) to teach or reinforce polygons, sides and angles and to explore the concept of greed with their students. The triangle is shown in its many important roles such as supporting bridges, acting as a sail or a roof and being slices of pie. But the triangle becomes bored and wants to change by having one more side and angle. The shapeshifter grants the triangle's wish and turns it into a quadrilateral. Now the triangle is shown as a computer screen, picture frames and a baseball diamond. As your students might predict, the quadrilateral is bored again and goes back to the shapeshifter for more angles and sides. Teachers can use this book to teach defining and non-defining attributes of shapes as well as provide opportunities for students to build shapes based on particular attributes. Have 10 toothpicks in resealable sandwich bags for students to use to build the shapes in the story.

Geometry 1.G

Reason with shapes and their attributes.

Distinguish between defining attributes (e.g., triangles are closed and three-sided) versus non-defining attributes (e.g., color, orientation, overall size); build and draw shapes to possess defining attributes.

As you read the book, engage children in discussion and activities by asking the following questions based on the *I Can* statements:

- Have you ever heard of the word greedy? How would you describe someone who is greedy? (They don't share. They want everything for themselves. They are not happy with what they have. They want what someone else has.) We are going to read a book about a greedy triangle. You will each receive a resealable sandwich bag of toothpicks. I want you to make a triangle with some of your toothpicks. How many sides does a triangle have? (3.) So take out only three toothpicks and make a triangle on your desk.

- I will also make a triangle on the document camera. Let's look at our triangle and see if we can think of some things that are shaped like a triangle. (The roof of a house, a piece of cake or pie or pizza, a sail for a sailboat, etc.) On the first page, we can see there are lots of things the triangle likes to do. We named some of these and there are even more!

- Now we see that the triangle is feeling "dissatisfied." What do you think this word means? Look at the illustration of the triangle to help you. (He is sad. He does not feel good.) He is unhappy and goes to the shapeshifter and asks him for two things that he thinks will make his life more interesting. He asks for one more side and one more angle. I want you to take out one toothpick and add a side to your three-sided figure. Now how many sides do you have? (4.) When you add a side to a shape, you also add an angle. An angle is like the corner of a shape.

- The special name for any four-sided shape is a *quadrilateral*. So let's look around the classroom right now and see if we can find shapes that have four sides. (There are windows, tables, desks, doors, computer screens, books, etc.) In the book, we can see that the quadrilateral is happy being the square in a checkerboard, a rectangular movie screen and computer screen, and many more things. But it is still the greedy triangle after all, so what do you predict it will do? (It will not be happy and will go to the shapeshifter again.)

- Before we see what the quadrilateral is going to do, I am going to have you solve a word problem that is based on this story. If you *can find a pattern in a problem* then you can solve it:

  The greedy triangle had 3 sides and 3 angles. Then it went to the shapeshifter and became a quadrilateral. The quadrilateral has 4 sides and 4 angles. Now it wants to go to the shapeshifter again to become a pentagon. How many sides and how many angles do you think it will get? Why do you think so?

- If we keep track of the number of sides and angles of the shapes in the book so far, it can help us *find a pattern in the problem*. I will make a table and you help me fill in the sections (Table 7.5.)

TABLE 7.5

Shape	Number of sides	Number of angles
triangle	3	3
quadrilateral	4	4
pentagon		

- First let's see if we can find a relationship between the number of sides and the number of angles from the table. (There are 3 sides and 3 angles. Then there are 4 sides and 4 angles.) So the number of sides is the same as the number of angles. We can look in the rows on the table and see that relationship. Now let's see if we can find a pattern in the table in the columns that can help us solve the problem. (There were 3 sides and angles, and then there were 4. The numbers are going up by ones.)

- So there is a pattern in the shapes that appear in the story. If there were 3 and then 4 sides and angles, how many do you think the quadrilateral will get from the shapeshifter when it becomes a pentagon? (5) Why do you think it will get 5 sides and 5 angles? (Because 5 comes after 4 and it is going in order. And if it will get 5 sides then it has to get 5 angles because those are always the same.)

- Now I'll read the next page and we'll see if we are correct. Yes, it does get one more side and one more angle so it now has 5 sides and 5 angles. I also want you to take out another toothpick so you can change your quadrilateral into a pentagon.

■ I think you are ready for another word problem based on this story. If you *can find a pattern in a problem* then you can solve it:

> The quadrilateral went to the shapeshifter to become a pentagon. Now it is dissatisfied again and wants to become a hexagon. How many sides and how many angles do you think it will get? Why do you think so?

■ If we continue to keep track of the number of sides and angles of the shapes in the book, it can help us *find a pattern in the problem*. Help me fill in the remaining sections (Table 7.6.)

TABLE 7.6

Shape	Number of sides	Number of angles
triangle	3	3
quadrilateral	4	4
pentagon	5	5
hexagon		

■ If we continue to look for a pattern in the columns, we can figure out the answer to the problem. If there were 3 and then 4 and then 5 sides and angles, how many do you think the pentagon will get from the shapeshifter when it becomes a hexagon? (6.) Why do you think it will get 6 sides and 6 angles? (Because 6 comes after 5 and it is going in order. And if it will get 6 sides then it has to get 6 angles because those are always the same.)

■ Now I'll read the next page and we'll see if we are correct. Yes, it does get one more side and one more angle so it now has 6 sides and 6 angles. I also want you to take out another toothpick so you can change your pentagon into a hexagon. We can also stop right now to think about how adding one more side also creates one more angle. Each time the triangle changed its shape, I asked you to add one more toothpick to make another side. But did you have to do anything to make another angle or did your shape automatically get another angle? (When we add another toothpick for a side, it makes another angle.) Yes, another angle is formed when another set of sides are joined together.

■ You might think the hexagon is finally happy but the book says, "the shape became restless, dissatisfied, and unhappy with its life." What do you think it will do now? (It will go to the shapeshifter again; it will get more sides and shapes; it will keep changing, etc.) You can see on this page that it is changing into more shapes such as a heptagon, an octagon, a nonagon and a decagon. If we fill in our chart we can see how many sides and angles are in those other shapes (Table 7.7).

TABLE 7.7

Shape	Number of sides	Number of angles
triangle	3	3
quadrilateral	4	4
pentagon	5	5
hexagon	6	6
heptagon	7	7
octagon	8	8
nonagon	9	9
decagon	10	10

Have your students use their toothpicks to create the heptagon and the rest of the shapes, noticing how another angle is formed as they add another side. Or you can create word problems based on those for the pentagon and hexagon so your students can practice explaining how they can use a pattern to solve the problem. As a follow-up activity, give students several cut-outs of each shape in the story as well as circles, ovals and figures that are not closed, varying in size and color. Include some polygons that are regular and some that are not, so they can see the sides are not always congruent. Then the students can sort the shapes according to attributes such as number of sides, number of angles, color, closed figures and shapes that do not have sides.

Concluding Remarks

Encourage your students to look for patterns in their environment but to also seek opportunities to use patterns in their problem solving. Whether it is in number and operations or in geometry, students should practice noticing patterns and ways to group elements. This is also a chance to introduce and reinforce mathematical terms so students have a way to discuss, describe and explain their patterns of structure.

8

Apply Repeated Reasoning

Unpacking the Standard—SMP 8: Look For and Express Regularity in Repeated Reasoning

Teachers should teach shortcuts with caution. Often students rely on the shortcut but are unable to explain why the shortcut works such as moving a decimal point to the right to multiply by 10. However, when children discover a shortcut on their own, they tend to understand the mathematics behind it. Teachers should explicitly teach students how to recognize opportunities in word problems for using a shortcut such as when a calculation is repeated. For example, if a word problem involves adding $5 + 5 + 5 + 5$, the students should stop and discuss how it would be more efficient to count by fives or to multiply instead of writing out this addition problem. In order to find shortcuts, students have to pay attention to the details of the problem as well as be aware of the goals of the problem. In the previous example, it does not matter how the problem solver arrived at the answer of 20, but that they understand there are efficient ways to arrive at the same answer as long as it satisfies the goal of the problem.

Children who are successful at SMP 8 are familiar with efficient strategies and shortcuts that can be applied to various word problems. They recognize the various types of problem-solving situations for the various operations (see Appendix) based on the structure of the problem and if calculations are repeated. They pay attention to the details of the problem while they keep the goal of the solution of the problem in mind.

In order to fully apply SMP 8 when approaching a word problem, children should be able to take ownership of their procedures by using the following *I Can* statements:

- *I can look for repeated calculations.*
- *I can create a shortcut.*
- *I can pay attention to details while I think about the goal of the problem.*

Early Problem Solvers

Bunches of Buttons: Counting by Tens by Michael Dahl—Counting to 100 by Tens

In this book, *Bunches of Buttons: Counting by Tens* (2006), Billy collects buttons in a jar, finding them in various places in his house. As he finds the buttons, he counts them in groups of ten, eventually collecting 100 buttons. In the illustrations of the collections of buttons, each set of ten is a different color in order to distinguish each set. On each page, the decades are also shown in numeral form with ten dots above each numeral, from 10–100. Your students can practice writing, representing and counting by tens.

Counting and Cardinality K.CC

Know number names and the count sequence.

Count to 100 by ones and by tens.

Write numbers from 0 to 20. Represent a number of objects with a written numeral 0–20 (with 0 representing a count of no objects).

As you read the book, engage children in discussion and activities by asking the following questions based on the *I Can* statements:

- In this book, Billy likes to collect buttons. Let's see how many he can find in his house. On page 3, how many buttons does he have in his jar? (0.) I want you to write the numeral 0 on your dry erase board. If he has *zero* buttons that means he has no buttons in his jar. Now you can erase the zero.

- On page 4 he finds ten red buttons. I want you to write the numeral 10 on your dry erase board. Which digits did you have to write to make ten? (We had to write a 1 and a 0.)

- You can erase the ten. Let's see how many buttons he finds on page 6. How many does he find and what color are the buttons? (He finds ten orange buttons.) He put the ten orange buttons in his jar with the ten red buttons. We are going to count the buttons in the jar. Ready? 1, 2 . . . 20. There are now 20 buttons in his jar. Could we have counted them another way? Let's see if we *can create a shortcut* for the next page. Write 20 on your dry erase board so we can be sure everyone is writing it correctly. Which digits did you have to write to make twenty? (We had to write a 2 and a 0.)

- You can erase the twenty. Let's see how many buttons he finds on page 8. How many does he find and what color are the buttons? (He finds ten purple buttons.) He put the ten purple buttons in his jar with the ten red buttons and the ten orange buttons. We are going to count the buttons in the jar.

Ready? 1, 2 . . . 30. There are now 30 buttons in his jar. Could we have counted them another way? I want to see if we *can create a shortcut* for this page because it's taking longer to count the buttons by ones. Is there a faster way to count the sets of buttons? (We can count them by tens.) Let's try it: 10, 20, 30. That was a great shortcut!

■ Write 30 on your dry erase board so we can be sure everyone is writing it correctly. Which digits did you have to write to make thirty? (We had to write a 3 and a 0.)

■ You can erase the thirty. Let's see how many buttons he finds on page 10. How many does he find and what color are the buttons? (He finds ten green buttons.) He put the ten green buttons in his jar with the ten red buttons, ten orange buttons and ten purple buttons. We are going to count the buttons in the jar. Ready? 1, 2 . . . 40. There are now 40 buttons in his jar. How can we count them with our shortcut? (We can count them by tens.) Let's try it: 10, 20, 30, 40.

■ Write 40 on your dry erase board so we can be sure everyone is writing it correctly. Which digits did you have to write to make forty? (We had to write a 4 and a 0.)

■ You can erase the forty. Let's see how many buttons he finds on page 12. How many does he find and what color are the buttons? (He finds ten pink buttons.) He put the ten pink buttons in his jar with the ten red buttons, ten orange buttons, ten purple buttons and ten green buttons. We are going to count the buttons in the jar. Ready? 1, 2 . . . 50. There are now 50 buttons in his jar. How can we count them with our shortcut? (We can count them by tens.) Let's try it: 10, 20, 30, 40, 50.

■ Write 50 on your dry erase board so we can be sure everyone is writing it correctly. Which digits did you have to write to make fifty? (We had to write a 5 and a 0).

■ You can erase the fifty. Let's see how many buttons he finds on page 14. How many does he find and what color are the buttons? (He finds ten blue buttons.) He put the ten blue buttons in his jar with the ten red buttons, ten orange buttons, ten purple buttons, ten green and ten pink buttons. We are going to count the buttons in the jar. Ready? 1, 2 . . . 60. There are now 60 buttons in his jar. How can we count them with our shortcut? (We can count them by tens.) Let's try it: 10, 20, 30, 40, 50, 60.

■ Write 60 on your dry erase board so we can be sure everyone is writing it correctly. Which digits did you have to write to make sixty? (We had to write a 6 and a 0).

■ You are getting very good at this so I'm going to give you a little word problem to solve:

Billy has 60 buttons in his jar. He finds 10 yellow buttons. Now how many buttons will he have in his jar?

■ In order to solve this problem, think about the counting pattern. He always finds the buttons in sets of ten so it is easy for him to count by tens. This is a type of *repeated calculation* because he adds ten more buttons in his jar every time. Who thinks they know the answer to the problem? (He will have 70 buttons in his jar.) How did you know? (If we count by tens we will say 70 next.)

■ You can erase the sixty. Let's check pages 16 and 17 to see if we are correct. He finds ten yellow buttons and puts them in the jar with the other 60 buttons. Let's count the buttons in the jar by tens, 10, 20 . . . 70. Can you write the numeral 70 on your dry erase board? Which digits did you have to write to make seventy? (We had to write a 7 and a 0.)

■ Here is another little word problem to solve but it has two parts to the solution:

Billy has 70 buttons in his jar. How many buttons do you think he will find next? How many buttons will he have in his jar altogether?

■ The first part of this word problem asks about the number of buttons he will find next. Think about the pattern in the book. How many buttons does he find each time? (10.) So we predict he will find 10 buttons. Now we can solve the second part of the problem. How many buttons will he have if he adds the 10 buttons to the 70 buttons he already has in his jar? (80.) How do you know? (If we count by tens we will say 80 next.) That's right. It is the same as adding 10 more which is the *repeated calculation* in the book. Counting by tens to get to 80 is *a shortcut* instead of counting by ones.

■ You can erase the seventy. Let's check pages 18 and 19 to see if we are correct. He finds ten white buttons and puts them in the jar with the other 70 buttons. Let's count the buttons in the jar by tens, 10, 20 . . . 80. Can you write the numeral 80 on your dry erase board? Which digits did you have to write to make eighty? (We had to write an 8 and a 0.)

■ We're almost done with the story. How many buttons do you think he will have in his jar on the next page? (90.) How do you know? (If he finds 10 more buttons then he will count by tens again. If we count by tens we will say 90 next.)

■ You can erase the eighty. Let's check pages 20 and 21 to see if we are correct. He finds ten brown buttons and puts them in the jar with the other 80 buttons. Let's count the buttons in the jar by tens, 10, 20 . . . 90. Can you write the numeral 90 on your dry erase board? Which digits did you have to write to make ninety? (We had to write a 9 and a 0.)

■ What if Billy keeps looking and finds another ten buttons in his house? How many buttons would he have if he put ten more in his jar of 90 buttons? Erase your 90 and write the answer on your dry erase board. (100.) Let's see if you are correct by looking at pages 22 and 23. He finds ten black buttons and adds them to his jar of 90 buttons. Let's count the buttons in the jar by tens, 10, 20 . . . 100!

Your students can practice writing all of the numerals from 0 to 100 by tens. They can count math manipulatives that are in sets of ten such as base 10 blocks and Unifix cubes, or collections of objects they brought from home. You can read the other books by the same author that are about counting by tens as well as his books about counting by fives and twos.

Developing Problem Solvers

How Big is a Foot? by Rolf Myller—Iterating Length Units

This book, *How Big is a Foot?* (1962), can be used to explore the need for standardizing the measurement of a foot. The King wants to get his wife, the Queen, a very special birthday present but she already owns everything. He decides to have a bed made for her because beds had not been invented yet so she did not own a bed. He speaks to his Prime Minister, who speaks to the Chief Carpenter, who speaks to the apprentice about how to make the bed. The King uses his feet to measure the Queen and gives the dimensions to the Prime Minister. But as the dimensions are conveyed to the apprentice, who is much smaller than the King, he uses his tiny feet to measure for the bed so that it turns out to be too small. Your students can practice measuring with their feet and other nonstandard units by laying them end to end.

Measurement and Data 1.MD

Measure lengths indirectly and by iterating length units.

Express the length of an object as a whole number of length units, by laying multiple copies of a shorter object (the length unit) end to end; understand that the length measurement of an object is the number of same-size length units that span it with no gaps or overlaps. *Limit to contexts where the object being measured is spanned by a whole number of length units with no gaps or overlaps.*

As you read the book, engage children in discussion and activities by asking the following questions based on the *I Can* statements:

■ Imagine if you had to buy a birthday gift for a queen who already owns everything. In this book, the King wants to give his wife, the Queen, a very special gift, something she does not already own. Let's find out what he decided to give to her.

■ So we find out that the King wants to give the Queen a bed. Why is that a good birthday gift for the Queen? (Because she doesn't have a bed; beds were not invented yet; she would be the first person to ever have a bed, etc.)

- Since beds were not yet invented, the King had to have a bed made. Who are all of the people involved in the process of making this bed? (The King, Prime Minister, Chief Carpenter, apprentice and the Queen) What do you also notice about the size of these men in the book? (The King is really big and the other men are shorter; the apprentice is very small.)

- What question did the apprentice need answered before he could start making the bed? (He needed to know how big to make the bed.) If no one had ever made a bed, he would need to know the size of the bed. Since the bed is for the Queen, what do you think the King will have to do to answer the apprentice's question? (The King will have to find out how big to make the bed; the King will measure the Queen.)

- In this story, rulers, tape measures and yardsticks have not been invented yet. So how could the King measure the Queen? (He could see how tall she is compared to him; He could see how much rope or string is the same size as the Queen; he could trace the Queen on a big piece of paper, etc.) Great ideas! Let's see how he measures the Queen.

- I will show you the page where the King has the Queen lay on the ground in her pajamas and her crown, because she sometimes wore it to sleep. Turn to your partner and explain what you think the King did to measure the Queen. (He walked around her to see how tall she is and how wide she is.)

- What did the King use as a tool to measure the Queen? (He used his feet.) How does he know how long to make the bed? (It took six of his feet to walk on the side of the Queen.) How does he know how wide to make the bed? (It took three of his feet to walk across the bottom of the Queen.)

- Look at the picture of the King's feet as he walked around the Queen. He made sure there were no spaces between his feet across the bottom and again when he walked along the side of the Queen. He also didn't overlap his feet but made sure they were end to end. Let's take a few steps like the King did, walking heel to toe, heel to toe.

- So it seems like the problem is solved. How big should the bed be to fit the Queen? (It should be 3 feet wide and 6 feet long.) If there are no rulers, tape measures or yardsticks, how will the apprentice know what 3 feet wide and 6 feet long looks like? (He can ask the King how big to make the feet; he can use his own feet.)

- I will show you the page where the apprentice uses his own feet to measure three feet wide and six feet wide. Does it look like the apprentice made sure there were no spaces between his feet when he measured? (Yes) Does it look like he walked heel to toe so his feet were end to end? (Yes) Do you think he will be able to make the bed the correct size so it will be big enough for the Queen? Discuss this with your partner.

- We can see that the apprentice made the bed too small for the Queen and was thrown in jail! There is a question here too, "Why was the bed too small for the Queen?" You can write a few sentences or draw a picture to answer the question. You will have to *pay attention to details while you think about the goal of the problem.* The goal is to answer why the bed was too small for the Queen but your answer will be based on the details of the problem.

■ What details did you have to think about or use to solve the problem? (We had to think about the size of the apprentice's feet and the size of the King's feet; we had to use details from the other parts of the book like the size of the apprentice who was much smaller than the King.)

■ It looks like the apprentice is trying to figure out the same problem while he sits in the jail. He realizes that he needs to know the size of the King's foot so the bed can be "three King's feet wide and six King's feet long." The King is very busy so the apprentice will have to *create a shortcut* to solve his problem. What can he do to find out the size of the King's foot so he can always make things for the King? (He can trace the King's foot; he can use string or rope and cut it the same size as the King's foot, etc.)

■ Well they ask the sculptor to make a copy of the King's foot and give it to the apprentice and now he can use the copy to make the bed. So in this kingdom, they will always use the King's foot whenever they measure anything. That is why we call a ruler a foot, which is exactly 12 inches long. We will always know the answer to the question, How big is a foot?

Your students can measure the length of various objects with their feet and see if they get the same answer. If you measure the same length with your feet, will the number of "feet" be more or less than their answer? Are there other body parts they can use to measure and compare, such as their pinky finger or their leg? Introduce your students to other ways that cultures have historically used their body to measure such as using paces for distances and hand spans for lengths.

Advanced Problem Solvers

The King's Commissioners by Aileen Friedman—Representing Tens and Ones

In this book, *The King's Commissioners* (1994), the King has so many royal commissioners that he constantly loses track of who is taking care of what. He decided to count them so he can begin to organize their duties. He asks his First and Second Royal Advisors to help count the commissioners as they come in the throne room. As they come in one at a time, the King tries to count them in his head while the Royal Advisors make tallies. When the King loses count and can't make sense of the tallies, his daughter asks the Royal Organizer to put all of the commissioners in rows of 10 so they will be easier to count. Your students can learn how to read and write two-digit numbers, organize quantities into tens with any leftovers counted as ones and group numbers by 2s and 5s.

Number and Operations in Base Ten 1.NBT

Understand place value.

Understand that the two digits of a two-digit number represent amounts of tens and ones. Understand the following as special cases:

a. 10 can be thought of as a bundle of ten ones—called a "ten."

b. The numbers from 11 to 19 are composed of a ten and one, two, three, four, five, six, seven, eight, or nine ones.

c. The numbers 10, 20, 30, 40, 50, 60, 70, 80, 90 refer to one, two, three, four, five, six, seven, eight, or nine tens (and 0 ones).

As you read the book, engage children in discussion and activities by asking the following questions based on the *I Can* statements:

- This story is about a King who has many, many royal commissioners who are in charge of things such as Flat Tires and Things That Go Bump in the Night. One day he decides to count them all so he can organize them. Let's see what happens when he tries to keep track of all of them as they come into his throne room.

- Since there are so many royal commissioners, the King asks his two Royal Advisors to help keep count. They each have a large notepad. What do you think they are writing on their notepad? (They are writing down numbers as the commissioners come in; they are adding up all of the people, etc.) It says in the book they are making tallies on the notepads. How did the Royal Advisors *create a shortcut* for counting all of the commissioners? (They can make a tally for each person that comes in, then count up the tally marks at the end.)

- The King is trying to count the commissioners but his daughter the Princess comes in and he loses count when he gives her a hug. When the King asks the First Royal Advisor how many commissioners he counted, he shows the King the notepad. It has tally marks on it for every commissioner who came in the door and then he circled the tally marks in twos. How could he figure out how many commissioners came in? (He could count them by twos.)

- The King is confused by the First Royal Advisor's method so he asks the Second Royal Advisor to show his notepad. He also made a tally mark for every commissioner but he circled the tally marks in groups of five. How could he figure out how many commissioners came in? (He could count them by fives.)

- The King is confused by the Second Royal Advisor's method. The Princess says she has a way to figure out how many commissions there are. She tells the Royal Organizer to put the commissioners in rows so there are 10 in each row. There were four rows of 10 and 7 commissioners left over. How could she figure out how many commissioners came in? (She could count them by tens and then add in the ones that are left over.)

- Let's see how The Princess figures it out. She walks down the rows and says, "10, 20, 30, 40, plus 7 more makes 47." How did she *create a shortcut* for counting the commissioners? (She put them in groups of 10 so she could count by tens. Then it was easy to add 7 more.)

- Good problem solvers not only *can create a shortcut* but they know they *can look for repeated calculations*. In order to figure out how many commissioners in all, which is the problem the King is trying to solve, the First Royal Advisor can look at his notepad and look for an opportunity to use a repeated calculation. That means he will do the same thing over and over. Since he circled the tallies by twos, what is his repeated calculation? (He can count by twos and then add in the leftover tally mark.) He counted by twos until he reached 46 and then added 1 more to make 47.

- Now the Second Royal Advisor can look at his notepad and look for an opportunity to use a repeated calculation. Since he circled the tallies by fives, what is his repeated calculation? (He can count by fives and then add in the two leftover tally marks.) He counted by fives until he reached 45 and then added 2 more to make 47.

You can use this book as a springboard for a discussion about looking for shortcuts in a problem, such as putting numbers into groups of twos, fives or tens. Show examples of how you can look for opportunities to repeat calculations in addition word problems or when counting up groups of numbers.

Concluding Remarks

Continue using examples of word problems that allow students to create shortcuts and look for repeated calculations that can help them arrive at their solution. Use a think-aloud to model how to read word problems for the details that can help with the solution process. Once your students are able to do this with one-step word problems, model how to look for and record details that can help solve two-step word problems.

Next Steps

Once your students have been exposed to the Standards for Mathematical Practice using some of the children's literature from this book, choose other books that can lend themselves to the SMP as well as to the content standards. Share ideas with your colleagues or start a list with other teachers in an online environment. Continue improving the culture of mathematical problem solving in your school by introducing grade-level or school-wide problem-solving situations such as figuring out the area and perimeter of the playground, graphing the amount of garbage generated by the school each day, or establishing a school-wide fundraiser conducted by the students.

Include a poster in your classroom with the eight SMP and the *I Can* statements so students can refer to them as they progress through problem-solving activities throughout the year. Students should be able to explain which SMP they are using and why. Share explanations of the problem-solving process with parents in newsletters, on the school or classroom website, at Open House or Parent-Teacher Conferences and even in the math work that students bring home. Create extensions of the classroom activities to be sent home so students can work with their family to solve problems using the SMP.

Encourage your school to begin housing resources for incorporating problem solving and the Common Core State Standards for Mathematics such as professional books, periodicals, DVDs, games, center activities and children's literature that can be used with math content areas. Collaborate with other teachers to find ways to integrate other content areas into math such as art, music and physical education. Students can study concepts in geometry while working on an art project, explore patterns and counting while they plan an instrument in music and measure distance and time in physical education class.

I hope this book has helped illuminate the eight SMP as well as some of the content standards through the use of children's literature. Perhaps this book served as an introduction to the language used in the Common Core State Standards for Mathematics document or provided ideas as to how the SMP can be broken down for younger students. Please persist in teaching your students the importance of the process of problem solving rather than simply focusing on which student can provide the correct answer.

Appendix

	Result unknown	Change unknown	Start unknown
Add to	Two bunnies sat on the grass. Three more bunnies hopped there. How many bunnies are on the grass now? $2 + 3 = n$	Two bunnies were sitting on the grass. Some more bunnies hopped there. Then there were five bunnies. How many bunnies hopped over to the first two? $2 + n = 5$	Some bunnies were sitting on the grass. Three more bunnies hopped there. Then there were five bunnies. How many bunnies were on the grass before? $n + 3 = 5$
Take from	Five apples were on the table. I ate two apples. How many apples are on the table now? $5 - 2 = n$	Five apples were on the table. I ate some apples. Then there were three apples. How many apples did I eat? $5 - n = 3$	Some apples were on the table. I ate two apples. Then there were three apples. How many apples were on the table before? $n - 2 = 3$ Total n

	Total unknown	Addend unknown	Both addends unknown
Put together/ Take apart	Three red apples and two green apples are on the table. How many apples are on the table? $3 + 2 = n$	Five apples are on the table. Three are red and the rest are green. How many apples are green? $3 + n = 5, 5 - 3 = n$	Grandma has five flowers. How many can she put in her red vase and how many in her blue vase? $5 = 0 + 5, 5 = 5 + 0$ $5 = 1 + 4, 5 = 4 + 1$ $5 = 2 + 3, 5 = 3 + 2g$

continued overleaf . . .

Continued

	Difference unknown	Bigger unknown	Smaller unknown
Compare	*("How many more?" version):* Lucy has two apples. Julie has five apples. How many more apples does Julie have than Lucy? *("How many fewer?" version):* Lucy has two apples. Julie has five apples. How many fewer apples does Lucy have than Julie? 2 + n = 5, 5−2 = n	*(Version with "more"):* Julie has three more apples than Lucy. Lucy has two apples. How many apples does Julie have? *(Version with "fewer"):* Lucy has 3 fewer apples than Julie. Lucy has two apples. How many apples does Julie have? 2 + 3 = n, 3 + 2 = n	*(Version with "more"):* Julie has three more apples than Lucy. Julie has five apples. How many apples does Lucy have? *(Version with "fewer"):* Lucy has 3 fewer apples than Julie. Julie has five apples. How many apples does Lucy have? 5−3 = n, n + 3 = 5

Source: Common addition and subtraction situations based on the Common Core State Standards for Mathematics.

References

Blaisdell, M. (2010). *If you were a quadrilateral*. Minneapolis, MN: Picture Window Books.

Burns, M. (1997). *Spaghetti and meatballs for all!* New York: Scholastic.

Burns, M. (1994). *The greedy triangle*. New York: Scholastic.

Carle, E. (1972). *Rooster's off to see the world*. New York: Simon & Schuster.

Common Core State Standards Initiative (CCSSI) (2010). *Common Core State Standards*. Washington, DC: National Governors Association Center for Best Practices and Council of Chief State School Officers.

Dahl, M. (2006). *Bunches of buttons: Counting by tens*. Minneapolis, MN: Picture Window Books.

Friedman, A. (1994). *The king's commissioners*. New York: Scholastic.

Garland, M. (2007). *How many mice?* New York: Dutton Children's Books.

Giganti, P. (1992). *Each orange had 8 slices*. New York: Greenwillow Books.

Hutchins, P. (1986). *The doorbell rang*. New York: Greenwillow Books.

Jonas, A. (1995). *Splash!* New York: Greenwillow Books.

Leedy, L. (1997). *Measuring Penny*. New York: Henry Holt & Co.

McNamara, M. (2007). *How many seeds in a pumpkin?* New York: Schwartz & Wade Books.

Murphy, S. (2006). *Mall mania*. New York: Harper Collins Publishers.

Murphy, S. (2004). *Earth Day-hooray!* New York: Harper Collins Publishers.

Murphy, S. (2002). *Bigger, better, best!* New York: Harper Collins Publishers.

Murphy, S. (1998). *Lemonade for sale*. New York: Harper Collins Publishers.

Myller, R. (1962). *How big is a foot?* New York: Dell Publishing.

Neuschwander, C. (2007). *Patterns in Peru: An adventure in patterning*. New York: Henry Holt & Co.

Neuschwander, C. (2005). *Mummy math: An adventure in geometry*. New York: Henry Holt & Co.

Reid, M. (1990). *The button box*. New York: Penguin Books.

Scieszka, J. (1995). *The math curse*. New York: Penguin Books.

Sturges, P. (1995). *Ten flashing fireflies*. New York: North-South Books.

Tang, G. (2003). *MATH-terpieces*. New York: Scholastic.

Viorst, J. (1978). *Alexander, who used to be rich last Sunday*. New York: Scholastic.

White, J. and Dauksas, L. (2012). The Common Core State Standards for Mathematics: Getting Started in K-grade 2. *Teaching Children Mathematics*, 18, 440–445.

Young, E. (1992). *Seven blind mice*. New York: Scholastic.